REVIEW OF THE RESEARCH PROGRAM OF THE **FreedomCAR** AND **Fuel Partnership**

THIRD REPORT

Committee on Review of the FreedomCAR and Fuel Research Program, Phase 3

Board on Energy and Environmental Systems

Division on Engineering and Physical Sciences

NATIONAL RESEARCH COUNCIL
OF THE NATIONAL ACADEMIES

THE NATIONAL ACADEMIES PRESS
Washington, D.C.
www.nap.edu

THE NATIONAL ACADEMIES PRESS 500 Fifth St., N.W. Washington, DC 20001

NOTICE: The project that is the subject of this report was approved by the Governing Board of the National Research Council, whose members are drawn from the councils of the National Academy of Sciences, the National Academy of Engineering, and the Institute of Medicine. The members of the committee responsible for the report were chosen for their special competences and with regard for appropriate balance.

This study was supported by Contract/Grant No. DE-AC26-08NT06206 between the National Academy of Sciences and the U.S. Department of Energy. Any opinions, findings, conclusions, or recommendations expressed in this publication are those of the author(s) and do not necessarily reflect the views of the organizations or agencies that provided support for the project.

International Standard Book Number-13: 978-0-309-15683-7
International Standard Book Number-10: 0-309-15683-1

Copies of this report are available in limited supply free of charge from the Board on Energy and Environmental Systems, National Research Council, 500 Fifth Street, N.W., Keck 943, Washington, DC 20001; (202) 334-3344

Additional copies of this report are available from the National Academies Press, 500 Fifth Street, N.W., Lockbox 285, Washington, DC 20055; (800) 624-6242 or (202) 334-3313 (in the Washington metropolitan area); Internet, http://www.nap.edu

Printed in the United States of America

THE NATIONAL ACADEMIES

Advisers to the Nation on Science, Engineering, and Medicine

The **National Academy of Sciences** is a private, nonprofit, self-perpetuating society of distinguished scholars engaged in scientific and engineering research, dedicated to the furtherance of science and technology and to their use for the general welfare. Upon the authority of the charter granted to it by the Congress in 1863, the Academy has a mandate that requires it to advise the federal government on scientific and technical matters. Dr. Ralph J. Cicerone is president of the National Academy of Sciences.

The **National Academy of Engineering** was established in 1964, under the charter of the National Academy of Sciences, as a parallel organization of outstanding engineers. It is autonomous in its administration and in the selection of its members, sharing with the National Academy of Sciences the responsibility for advising the federal government. The National Academy of Engineering also sponsors engineering programs aimed at meeting national needs, encourages education and research, and recognizes the superior achievements of engineers. Dr. Charles M. Vest is president of the National Academy of Engineering.

The **Institute of Medicine** was established in 1970 by the National Academy of Sciences to secure the services of eminent members of appropriate professions in the examination of policy matters pertaining to the health of the public. The Institute acts under the responsibility given to the National Academy of Sciences by its congressional charter to be an adviser to the federal government and, upon its own initiative, to identify issues of medical care, research, and education. Dr. Harvey V. Fineberg is president of the Institute of Medicine.

The **National Research Council** was organized by the National Academy of Sciences in 1916 to associate the broad community of science and technology with the Academy's purposes of furthering knowledge and advising the federal government. Functioning in accordance with general policies determined by the Academy, the Council has become the principal operating agency of both the National Academy of Sciences and the National Academy of Engineering in providing services to the government, the public, and the scientific and engineering communities. The Council is administered jointly by both Academies and the Institute of Medicine. Dr. Ralph J. Cicerone and Dr. Charles M. Vest are chair and vice chair, respectively, of the National Research Council.

www.national-academies.org

COMMITTEE ON REVIEW OF THE FREEDOMCAR AND FUEL RESEARCH PROGRAM, PHASE 3

VERNON P. ROAN, *Chair,* University of Florida (professor emeritus), Gainesville
DEBORAH LYNN BLEVISS, Independent Consultant, Falls Church, Virginia
DAVID L. BODDE, Clemson University, South Carolina
KATHRYN BULLOCK, Coolohm, Inc., Blue Bell, Pennsylvania
HARRY E. COOK, NAE,[1] University of Illinois (professor emeritus)
GLENN A. EISMAN, Rensselaer Polytechnic Institute, Troy, New York
W. ROBERT EPPERLY, Consultant, Mountain View, California
WILLIAM D. ERNST, EnerSysCon, Troy, New York
DAVID E. FOSTER, University of Wisconsin, Madison
GERALD GABRIELSE, NAS,[2] Harvard University, Cambridge, Massachusetts
LINOS JACOVIDES, Delphi Research Labs (retired), Grosse Pointe, Michigan
HAROLD H. KUNG, Northwestern University, Evanston, Illinois
CHRISTOPHER L. MAGEE, NAE, Massachusetts Institute of Technology, Cambridge
CRAIG MARKS, NAE,[3] Altarum, Bloomfield Hills, Michigan
GENE NEMANICH, Chevron Hydrogen Systems (retired), Scottsdale, Arizona
BERNARD ROBERTSON, NAE, DaimlerChrysler Corporation (retired), Bloomfield Hills, Michigan
R. RHOADS STEPHENSON, NASA Jet Propulsion Laboratory (retired), Consultant, La Cañada, California
KATHLEEN C. TAYLOR, NAE, General Motors Corporation (retired), Falmouth, Massachusetts
BRIJESH VYAS, LGS Bell Labs Innovations, Florham Park, New Jersey
ERIC WILLIAMS, Arizona State University, Tempe

[1] National Academy of Engineering.
[2] National Academy of Sciences.
[3] Deceased.

Subgroup on Systems Analysis and Simulation

Bernard Robertson, *Lead*
David L. Bodde
Harry E. Cook
David E. Foster
Harold H. Kung
Christopher L. Magee
Eric Williams

Subgroup on Advanced Combustion Engines and Emissions Control

David E. Foster, *Lead*
Deborah Lynn Bleviss
Harold H. Kung
Bernard Robertson
Kathleen C. Taylor

Subgroup on Electrochemical Energy Storage

Brijesh Vyas, *Lead*
Kathryn Bullock
Gerald Gabrielse
Linos Jacovides
Christopher L. Magee

Subgroup on Fuel Cells

Glenn A. Eisman, *Lead*
William D. Ernst
Gene Nemanich
Kathleen C. Taylor
Brijesh Vyas

Subgroup on Electric Propulsion, Electrical ystems and Power Electronics

Linos Jacovides, *Lead*
Kathryn Bullock
Harry E. Cook
William D. Ernst
Brijesh Vyas

Subgroup on Materials and Supplier Issues

Harry E. Cook, *Lead*
Deborah Lynn Bleviss
Glenn A. Eisman
W. Robert Epperly
Christopher L. Magee

Subgroup on Hydrogen Production and Delivery (including Off-Board Storage)

W. Robert Epperly, *Lead*
David L. Bodde
Glenn A. Eisman
Harold H. Kung
Gene Nemanich
R. Rhoads Stephenson
Eric Williams

Subgroup on Onboard Hydrogen Storage

Kathleen C. Taylor, *Lead*
William D. Ernst
Gene Nemanich
Christopher L. Magee
R. Rhoads Stephenson

Subgroup on Safety

R. Rhoads Stephenson, *Lead*
David L. Bodde
Harry E. Cook
W. Robert Epperly
Christopher L. Magee

Subgroup on Biofuels

Gene Nemanich, *Lead*
Deborah Lynn Bleviss
David L. Bodde
David E. Foster
Gerald Gabrielse
Gene Nemanich
Eric Williams

Subgroup on Electric Grid/Vehicle Charging Issues

David Bodde and Kathryn Bullock, *Leads*
Linos Jacovides
Brijesh Vyas

Project Staff

JAMES ZUCCHETTO, Director, Board on Energy and Environmental Systems
LaNITA JONES, Administrative Coordinator
DANA CAINES, Financial Associate

---

[1]National Academy of Engineering.
[2]National Academy of Sciences.

Preface

The FreedomCAR and Fuel Partnership has undergone several changes since its formation in January 2002. Initially, the Partnership was between the U.S. government (primarily the U.S. Department of Energy [DOE]) and the U.S. Council for Automotive Research (USCAR), whose members are Chrysler LLC, the Ford Motor Company, and General Motors Company. Soon after its inception, in September 2003 five energy companies were added as members: BP America, Chevron Corporation, ConocoPhillips, ExxonMobil Corporation, and Shell Hydrogen (U.S.). More recently, in 2008, two major power companies, DTE Energy (Detroit) and Southern California Edison, were added as members.

The Partnership developed a roadmap including many individual milestones and technical targets to pursue the original goal of "a full spectrum of vehicles that can operate free of petroleum and harmful emissions while sustaining the driving public's freedom of mobility and freedom of vehicle choice."[1] The long-term emphasis was on hydrogen fuel cell vehicles with hydrogen as the primary transportation fuel, but the Partnership envisioned utilizing transition technologies of advanced internal combustion engine vehicles and advanced hybrid electric vehicles en route to hydrogen/fuel cell vehicles.

With the change from the Bush to the Obama administration, there was an increase in emphasis on nearer-term technologies, especially those involving more electrification of the vehicles, such as plug-in hybrid electric vehicles. However, the charge to the Committee on Review of the FreedomCAR and Fuel Research Program, as well as presentations to the committee, involved performing an evaluation of activities between Phases 2 and 3, which included few activities

[1] See http://www1.eere.energy.gov/hydrogenandfuelcells/mypp/pdfs/introduction.pdf.

involving all-electric (battery electric vehicle or plug-in hybrid electric vehicle) technologies, or biofuels.

This report is the final full report, following a shorter letter report issued in July 2009,[2] for Phase 3 of the study of the Committee on Review of the FreedomCAR and Fuel Research Program as chartered by the National Research Council in the fall of 2008. It provides an overview of the structure and management of the Partnership as well as a discussion of the Partnership's adequacy, progress, and technical problem areas. Recommendations are also included in areas where the committee believes that improvements can be made.

Vernon P. Roan, *Chair*
Committee on Review of the FreedomCAR and Fuel Research Program

[2] See Appendix B in this report.

Acknowledgments

The committee wishes to thank the members of the FreedomCAR and Fuel Partnership, all of whom contributed a significant amount of their time and effort to this National Research Council (NRC) study by giving presentations at meetings, responding to requests for information, or providing valuable information. The chair also recognizes the committee members and the staff of the Board on Energy and Environmental Systems for their hard work in organizing and planning committee meetings and their individual efforts in gathering information and writing sections of the report.

This report has been reviewed in draft form by individuals chosen for their diverse perspectives and technical expertise, in accordance with procedures approved by the NRC's Report Review Committee. The purpose of this independent review is to provide candid and critical comments that will assist the institution in making its published report as sound as possible and to ensure that the report meets institutional standards for objectivity, evidence, and responsiveness to the study charge. The review comments and draft manuscript remain confidential to protect the integrity of the deliberative process. We wish to thank the following individuals for their review of this report:

Paul Blumberg, NAE, Independent Consultant, and Ford Motor Company (retired),
Jay Hakes, Director, Jimmy Carter Presidential Library and Museum,
Trevor Jones, NAE, ElectroSonics Medical, Inc.,
Fritz Kalhammer, Electric Power Research Institute (retired),
John Kassakian, NAE, Massachusetts Institute of Technology,
Jerome Milgram, NAE, Massachusetts Institute of Technology,

John Scott Newman, NAE, University of California, Berkeley,
William F. Powers, NAE, Ford Motor Company (retired),
Michael P. Ramage, NAE, ExxonMobil Research and Engineering Company (retired),
Daniel Sperling, University of California, Davis, and
Richard Teets, Delphi Research Laboratories (retired).

Although the reviewers listed above have provided many constructive comments and suggestions, they were not asked to endorse the conclusions or recommendations, nor did they see the final draft of the report before its release. The review of this report was overseen by Stephen Berry, NAS, University of Chicago. Appointed by the National Research Council, he was responsible for making certain that an independent examination of this report was carried out in accordance with institutional procedures and that all review comments were carefully considered. Responsibility for the final content of this report rests entirely with the authoring committee and the institution.

Contents

Tables, Figures, and Boxes

TABLES

FIGURES

BOXES

DEDICATION

This report is dedicated to our dear friend and colleague Dr. Craig Marks (1929-2009), who served on all three National Research Council (NRC) committee reviews of the FreedomCAR and Fuel Partnership, chairing the first two. Craig devoted most of his working career to advancing automotive technologies and generously volunteered his considerable knowledge and skills to work not only with this committee but also with all seven phases of its predecessor, the NRC committee that reviewed the Partnership for a New Generation of Vehicles (PNGV) program. His many contributions, friendly manner, and thoughtful insights will be greatly missed.

Summary

This report by the National Research Council's (NRC's) Committee on Review of the FreedomCAR and Fuel Research Program, Phase 3, is the third NRC review. The Phase 1 and Phase 2 reviews were issued in 2005 and 2008, respectively (NRC, 2005, 2008). The long-range goals of the Partnership focus on a transition to a highway transportation system that uses sustainable energy resources and reduces emissions, including net carbon emissions, on a life-cycle or well (source)-to-wheels basis (DOE, 2004). The Partnership focuses on pre-competitive research and development (R&D) that can help to accelerate the emergence of technologies that can meet the long-range goals.

The transition is envisioned by the Partnership to begin with the internal combustion engine (ICE)-powered light-duty vehicles that, because R&D leads to a better understanding of the in-cylinder combustion process, achieve increased efficiency and decreased emissions. It would continue with R&D leading to improved hybrid vehicles, both conventional hybrid electric vehicles (HEVs) and plug-in hybrid electric vehicles (PHEVs), while hydrogen and automotive fuel cell research also continues on the path leading to private-sector commercialization decisions by the year 2015. Increasing the capabilities of high-energy batteries for PHEVs could also lead to the market penetration of all-electric or battery electric vehicles (BEVs).

At the request of the U.S. Department of Energy (DOE), the committee issued an interim letter report (see Appendix B) in July 2009, about the time that the DOE fiscal year (FY) 2010 budget request to Congress essentially "zeroed out" the hydrogen and automotive fuel cell portions of the program in favor of developing nearer-term technologies. The letter report generally agreed with increased efforts on potentially nearer-term technologies in order to reduce petroleum dependence—efforts such as biofuels and the shifting of some transportation energy to the electrical grid (through the development of PHEVs and BEVs), but it expressed concern about effectively abandoning the longer-term hydrogen and automotive fuel cell programs. Given the uncertainty of technical and market success of many of the technologies under development, the committee believes that longer-term hydrogen and automotive fuel cell programs should remain in a balanced R&D portfolio of different options and is an appropriate strategy for the Partnership to pursue.

Since the DOE budget request for little or no funding (which was subsequently mostly reinstated by Congress) for hydrogen and automotive fuel cell R&D came *after* most of the accomplishments between Phases 2 and 3 of this study, this report focuses primarily on those accomplishments and on significant remaining barriers. Indeed, PHEVs, BEVs, and biofuels were not included in the initial FreedomCAR and Fuel Partnership program, so there were few such activities to evaluate or compare between Phases 2 and 3. The accomplishments were made possible, for the most part, by funding from the DOE, matching contributions from the DOE contractors, and efforts by the light-duty-vehicle original equipment manufacturers (OEMs)—the automotive manufacturers and their suppliers. As had previously been true, considerable guidance for needed research was provided by joint industry/government technical teams. This structure has been demonstrated to be an effective means of identifying high-priority, long-term precompetitive research needs while also addressing societal needs such as reducing petroleum dependence and greenhouse gas production. However, there are several very substantial barriers remaining that could inhibit positive fuel cell vehicle commercialization decisions by 2015.

Even though there had been considerable emphasis on hydrogen fuel and automotive fuel cells, there are a number of technical areas where R&D as well as technology validation programs have been pursued, including the following (see Chapter 3):

- ICEs potentially operating on conventional and various alternative fuels,
- Automotive and non-automotive fuel cell power systems,
- Hydrogen storage (especially onboard vehicles) systems,
- Electrochemical energy storage,
- Electric propulsion systems,
- Hydrogen production and delivery, and
- Materials leading to vehicle weight reductions.

In each of these technology areas, there are specific research goals (targets) established by the Partnership for 2010 and 2015. Program oversight is provided by an Executive Steering Group consisting of the DOE Assistant Secretary for Energy Efficiency and Renewable Energy (EERE) and a vice-presidential-level executive from each of the Partnership companies. The DOE EERE efforts are divided between the Vehicle Technologies (VT) program and the Hydrogen, Fuel Cells and Infrastructure Technologies program (HFCIT; the latter has been renamed the Fuel Cell Technologies [FCT] program). The Partnership collaborates with other DOE offices outside of EERE—for example, Fossil Energy, Nuclear Energy, Electricity Delivery and Reliability, and Science—as well as within EERE, such as the Biomass program, which are not part of the Partnership. The U.S. Department of Transportation (DOT) is also involved in safety-related activities as well as existing or new hydrogen (or other fuels) pipelines and delivery trucks, including those for hydrogen and biofuels.

The scope of this review, as for the previous reviews, is to assess the progress in each of the technical areas, comment on the overall adequacy and balance, and make recommendations, depending on issues identified by the committee, that will help the Partnership to meet its goals (see Chapter 1 for the committee's full statement of task). This Summary provides overall comments and a brief discussion of the technical areas covered more completely in the report and presents the committee's main conclusions and recommendations.[1] Additional recommendations appear in appropriate topic areas of Chapters 2 through 4.

OVERALL COMMENTS

Since the creation of the FreedomCAR program in January 2002, it has undergone significant changes in Partnership members, with five energy companies added in September 2003 and two electrical power companies in 2008. Even though the technologies involved are not all under the FreedomCAR and Fuel Partnership umbrella, the potential pathways to the long-term objectives of reduced petroleum consumption as well as reduced criteria emissions and reduced greenhouse gases (GHGs) seem also to have broadened. In the collective opinion of the committee, there are essentially three primary alternative pathways:

- Improved ICE vehicles coupled with greater use of biofuels,
- A shifting of significant portions of transportation energy from petroleum to the grid through the expanded use of PHEVs and BEVs, and
- The transition to hydrogen as a major transportation fuel utilized in fuel cell vehicles.

In general, the committee believes that the Partnership is effective in progressing toward its goals. There is evidence of solid progress in essentially all areas, even though substantial barriers remain (see Chapter 5).

Most of the remaining barriers relate to cost (e.g., fuel cells, batteries, etc.), although there are also substantial performance barriers (e.g., onboard hydrogen

[1] The numbering of the recommendations in this Summary matches that of the same recommendations as they appear in the text of the chapters.

storage, demonstrated fuel cell durability, adequate battery energy storage capability, etc.) and production and infrastructure barriers (e.g., the need for widespread affordable hydrogen if mass-produced fuel cell vehicles are to become a reality, a feedstock/production combination for biofuels that does not compete with food crops, etc.).

The fuel cell/hydrogen R&D is viewed by the committee as long-term, high-risk, high-payoff R&D that the committee considers not only to be appropriate, but also to be of the type that much of it probably would not get done without government support. Especially under the present economic conditions, the committee considers R&D for other precompetitive technologies, which could help reduce industry development times, also to be appropriate.

TECHNICAL AREAS

Advanced Internal Combustion Engines and Emission Controls

There seems to be little doubt that, regardless of the success of any of the pathways discussed, the ICE will be the dominant prime mover for light-duty vehicles for many years, probably decades. Thus, it is clearly important to perform R&D to provide a better understanding of the fundamental processes affecting engine efficiency and the production of undesirable emissions. Consequently, it is important to maintain an active ICE and liquid fuels R&D program at all levels, namely, in industry, government laboratories, and academia, to expand the knowledge base to enable the development of technologies that can reduce the fuel consumption of transportation systems powered by ICEs. This is the focus of the advanced combustion and emission control (ACEC) technical team.

All aspects of ICE operation are being pursued, and good progress is being made. Improvements have been achieved in ICE efficiency, including that for a hydrogen-fueled ICE, as well as advancements in different combustion regimes under investigation. With the projected increased use of biofuels, the technical team is now also engaged in fundamental combustion, emission, and kinetic studies of fuel derived from biomass. This work is aimed at understanding the fundamental changes that occur in ignition and emission-formation processes when different compounds, such as methyl esters that are found in biofuels, are used in the engine.

Recommendation 3-3. The advanced combustion and emission control technical team should engage with the biofuels research community to ensure that the biofuels research which the team is conducting is consistent with and leverages the latest developments in the field of biofuels R&D.

Fuel Cells

From the beginning of the FreedomCAR program, fuel cells have been a long-term focus. They, along with the hydrogen fuel that they would consume, offer the promise of zero emissions (produced directly by the vehicle[2]), high efficiency, and the smooth, quiet operation that goes with an electric propulsion system. With this focus, progress has been significant, with continuing increases in performance and decreases in projected costs essentially every year. However, in spite of the significant progress, no single fuel cell technology has attained the combination of performance and projected costs to be competitive with conventional systems.

With regard to the performance, planning, and management of fuel cell R&D, the committee's assessment is that the fuel cell technical team is well coordinated and is aligned with respect to the achievement of the goals and the longer-term, high-risk technology challenges, especially as the automotive OEMs are now road testing prototype fuel cell vehicles. Most performance targets have been met in various demonstrations but not with a single technology.

Key achievements highlighted by the DOE and made since the previous review are primarily performance- and cost-related. Demonstrated stack lifetimes in on-road vehicles have increased from approximately 1,250 hours to 1,977 hours. With the goal of 5,000 hours, this change represents a significant achievement since the Phase 2 NRC review. Furthermore, single-cell and short-stack tests at the laboratory scale have demonstrated (using accelerated test protocols) much longer run times (3M Company, 7,200 hours) that meet or exceed the goals of the Partnership.

Two separate DOE-funded studies, with independent oversight, have concluded that at volumes of 500,000 units produced per year, the cost per kilowatt for the fuel cell subsystem, including the fuel cell and the balance of plant, will be approximately \$60-\$70/kW for an 80 kW unit. These figures are still more than two times higher than the target, but significantly lower than the \$107/kW presented during the Phase 2 review (in 2008). The projected cost is split nearly evenly between the stack and the balance of plant.

The barriers that remain are both programmatic and technical. Programmatic issues relate to the coordination and execution of the high-risk research so that the solicitation timing and content address updated requirements of the Partnership. Technical barriers that still remain for the fuel cell stack are membrane and electrode life, in addition to cost. Both areas must remain the focus of the next round of solicitations.

---

[2] The total full-fuel-cycle emissions from either hydrogen-fueled vehicles or full battery electric vehicles depend on how the hydrogen or electricity is produced. Both types of vehicles would have the potential of zero emissions, but this would depend critically on the future technologies deployed for hydrogen or electricity production.

Recommendation 3-7. The DOE should establish backup technology paths, in particular for stack operation modes and stack components, with the fuel cell technical team to address the case of current technology selections determined not likely to meet the targets. The DOE should assess which critical technology development efforts are not yielding sufficient progress and ensure that adequate levels of support for alternative pathways are in place.

Onboard Hydrogen Storage

Onboard hydrogen storage is a key enabler for fuel-cell-powered vehicles. The primary focus of the hydrogen storage program is to foster the development and demonstration of commercially viable hydrogen storage technologies for transportation and stationary applications. A specific goal of the program is a vehicle driving range of greater than 300 miles between refuelings while simultaneously meeting vehicle packaging, weight, cost, and performance requirements. The program also includes life-cycle issues, energy efficiencies, safety, and the environmental impact of the applied hydrogen storage technologies.

Most of the work of the onboard hydrogen storage program is organized in four centers of excellence (COEs): the Metal Hydrides COE, the Chemical Hydrogen Storage COE, the Hydrogen Sorption Materials COE, and the Hydrogen Storage Engineering COE. The hydrogen storage technical team provides input to the DOE that guides the work of the COEs.

The physical storage of hydrogen on vehicles as compressed gas (and to a lesser extent liquid hydrogen) has emerged as the technology path for the early introduction of fuel cell vehicles. The hydrogen storage capacity of tanks is performance limiting for some vehicle architectures and is expensive, but it will not prevent vehicle introduction into the market. The storage capacity of current high-pressure tanks does not meet the long-term program goals but may be adequate for some applications for which the cost can be justified.

Research aimed at significantly higher hydrogen storage capability needs to be maintained as a primary research focus. Materials-based storage at the level required to meet all program targets is considered theoretically achievable, yet no single material has been identified that simultaneously meets all of the targets (weight, volume, efficiency, cost, packaging, safety, refueling ability, etc.). The discovery and development of materials for effective onboard hydrogen storage is high-technical-risk R&D not likely to be accomplished without continued research attention and government funding.

Recommendation 3-12. The hydrogen storage program is one of the most critical parts of the hydrogen/fuel cell vehicle part of the FreedomCAR and Fuel Partnership—both for physical (compressed gas) and for materials storage. It should continue to be funded, especially the systems-level work in the Hydrogen

Storage Engineering COE. Efforts should also be directed to compressed-gas storage to help achieve weight and cost reduction while maintaining safety.

Recommendation 3-15. The search for suitable onboard hydrogen storage materials has been broadly based, and significant progress is reported. Nonetheless the current materials are not close to the long-range goals of the Partnership. Onboard hydrogen storage R&D risks losing out to near-term applications for future emphasis and funding. The management of a long-term/short-term joint portfolio should be given consideration.

Electrochemical Energy Storage

Improved electrochemical energy storage technologies, especially batteries and ultracapacitors, are critical to the advancement of both the Partnership's nearer-term and long-term goals: significant improvement in their performance can result in greater electrification of vehicles (e.g., PHEVs and BEVs). These technologies have taken on even greater importance in the past year due to the priorities of the new administration seeking to achieve 1 million PHEVs on the road by 2015. The Partnership's budget for electrochemical energy technologies has increased as the importance of PHEV battery development has increased. At present, about 75 percent of the funding is focused on near- and midterm development efforts directed at HEV and PHEV applications, and only 25 percent is directed to long-term R&D. The Partnership should also take the initiative to strengthen its focus on longer-term research on high-energy batteries and the establishment of a path toward BEVs.

Lithium-ion (Li-ion) battery technologies hold promise of achieving the long-term goals of high power, energy, and other performance requirements for HEV and PHEV applications at anticipated costs lower than those for other battery systems. Thus, the Partnership is correctly focused on the development of these technologies while it continues to benchmark competing battery technologies and encourages research on higher-energy chemistries for BEV applications. At present, none of the Li-ion battery chemistries meets the combination of performance, life, and cost goals for 2012 PHEV requirements. Although significant progress has been recorded in the Li-ion battery performance, durability, and safety, there has been no significant reduction in the projected cost of batteries. The system battery cost for a production of 100,000 units per year for the HEV application remains at more than $900, almost twice the 2010 target of $500. Battery cost will play an even bigger role in the eventual success of the PHEV and BEV applications because much larger batteries are required.

Recommendation 3-17. The Partnership should significantly intensify its efforts to develop improved materials and systems for high-energy batteries for both plug-in electric vehicles and battery electric vehicles.

Recommendation 3-18. The Partnership should conduct a study to determine the cost of recycling batteries and the potential of savings from recycled materials. A research program on improved processes for recycling advanced batteries should be initiated in order to reduce the cost of the processes and recover useful materials and to reduce potentially hazardous toxic waste and, if necessary, to explore and develop new processes that preserve and recycle a much larger portion of the battery values.

Electric Propulsion and Electrical Systems

Electric propulsion is needed for HEVs, PHEVs, fuel cell vehicles (FCVs), and BEVs. In all of these cases the systems used can be distinguished by the size and power required as well as by the architecture. In addition to the prime mover (engine, fuel cell, or battery), the essential elements of the electric propulsion system are power electronics and one or two electrical machines. The power electronics converts the direct current provided by the fuel cell, the engine-driven generator, or the battery into an alternating current to power motors and wheels. The Partnership has appropriately focused on key technical areas that are precompetitive, with the objective of long-term reductions in size (volume and weight) and cost. To accomplish this, emphasis has been on better packaging, cooling, materials, and devices.

To achieve better performance while operating power electronics at higher temperatures, materials and new designs are being incorporated into devices to replace currently used materials. For example, silicon carbide (SiC) devices are being investigated, including approaches to reducing their costs. SiC can potentially permit much higher power density because devices can operate up to much higher temperatures. Higher power densities mean more compact devices, thus less materials and potentially lower cost, and the Partnership is investigating a new process for making SiC on silicon (Si) substrates. Building these devices on Si is a desirable first step.

In all of the electric drive vehicle configurations, at least one electric motor provides the power to drive the wheels, but in some cases an electrical generator is also needed. The machines are basically of two types—permanent magnet brushless motors and induction motors—and each has advantages and disadvantages. Permanent magnet motors are currently used in essentially all electric and hybrid vehicles (the only exception being the Tesla Roadster) because of their high efficiency. However, the materials utilized come from only a few places on Earth and are relatively expensive. Induction motors, by contrast, use common materials, are used widely in industrial applications, cost less, and could be advantageous if the costs of batteries decline enough so that the premium on motor efficiency becomes less important than motor cost.

Onboard battery charging during regenerative braking affects the efficiency and cost of the motor. As new battery and motor materials are developed, use of the Powertrain Systems Analysis Toolkit developed at the Argonne National Laboratory under DOE sponsorship may help quantify material cost and performance trade-offs between motor efficiencies and battery-charging requirements.

Recommendation 3-20. The Partnership should conduct a project to evaluate the effect of battery charging on lithium-ion battery packs as a function of the cell chemistries, cell geometries, and configurations in the pack; battery string voltages; and numbers of parallel strings. A standardized method for these evaluations should be developed to ensure the safety of battery packs during vehicle operation as well as during plug-in charging.

Recommendation 3-21. The Partnership should consider conducting a project to investigate induction motors as replacements for the permanent magnet motors now almost universally used for electric propulsion.

Structural Materials

The challenge to the materials technical team is to generate a cost-neutral 50 percent vehicle weight reduction. The 50 percent weight reduction is critical to reaching FreedomCAR goals for energy consumption and emissions. However, the target of no cost penalty for such a large weight reduction was unrealistic when set, and it remains unrealistic. A similar conclusion was stated in the Phase 2 report. What is missing at this juncture is a projection of what the cost penalty will likely be.

The target for a project on magnesium power-train components was to replace aluminum components with magnesium for a minimum weight savings of 15 percent and a cost penalty of less than $2.00 for each pound saved. This project was completed and exceeded the weight savings goal with a cost penalty of $3.00/lb at current magnesium prices. Although over the cost target, the outcome was judged as demonstrating that magnesium was both technically feasible and potentially cost-effective in these applications.

Cost has been a limiting factor in the use of commercial carbon-fiber-reinforced polymers for the design of automotive structures and body panels. As a result of one of the projects, it appears that major cost savings could be achieved through the use of polyolefin for the feedstock in many carbon fibers. The recycling of carbon-reinforced composites could also aid in the adoption of such materials while possibly helping to reduce costs.

Recommendation 3-22. The materials technical team should develop a systems-analysis methodology to determine the currently most cost-effective way for achieving a 50 percent weight reduction for hybrid and fuel cell vehicles. The materials team needs to evaluate how the cost penalty changes as a function of the percent weight reduction, assuming that the most effective mix of materials is used at each step in the weight-reduction process. The analysis should be updated on a regular basis as the cost structures change as a result of process research breakthroughs and commercial developments.

Recommendation 3-24. Methods for the recycling of carbon-reinforced composites need to be developed.

Hydrogen and Other Fuel/Vehicle Pathways

The Partnership in DOE's Office of Energy Efficiency and Renewable Energy includes the hydrogen production, delivery, and dispensing program, which is, in turn, part of the Fuel Cell Technologies program, which is within the EERE. The Fuel Cell Technologies program addresses a variety of means of producing hydrogen in distributed and centralized plants using technologies that can be made available in the short, medium, and long term. Three fuel technical teams are addressing these issues: fuel pathway integration, hydrogen production, and hydrogen delivery.

The hydrogen fuel/vehicle pathway integration effort is charged with looking across the full hydrogen supply chain from well (source) to tank. Specifically, the goals of this integration effort are to (1) analyze issues associated with complete hydrogen production, distribution, and dispensing pathways; (2) provide input to the Partnership on goals for individual components; (3) provide input to the Partnership on needs and gaps in the hydrogen analysis program including the important industrial perspective; and (4) foster full transparency in all analyses, including an independent assessment of information and analyses from other technical teams.

The DOE continues to make important progress toward understanding and preparing for the transition to hydrogen fuel. In the continuing source-to-wheels analyses, seven pathways, including both distributed and centralized hydrogen production, have been assessed, and the key drivers for pathway costs, energy use, and emissions have been identified.

Technology is available to produce and distribute hydrogen commercially, but it is not yet completely optimized or cost-effective for supplying local fueling stations. Research efforts are focused on (1) the further development of options that reduce cost, (2) reducing dependence on imported petroleum and natural gas, and (3) reducing greenhouse emissions.

As indicated above, this effort has thus far been focused on hydrogen. However, the Partnership is now examining three power system approaches, only one of which involves hydrogen: fuel cells powered by hydrogen, advanced combustion engines powered by biofuels, and PHEVs and BEVs powered by electricity. Clearly, additional effort is needed to develop meaningful comparisons of the fuel implications of these three approaches.

Recommendation 4-1. The DOE should broaden the role of the fuel pathways integration technical team (FPITT) to include an investigation of the pathways to provide energy for all three approaches currently included in the Partnership. This broader role could include not only the current technical subgroups for hydrogen, but also subgroups on biofuels utilization in advanced internal combustion engines and electricity generation requirements for PHEVs and BEVs, with appropriate industrial representation on each. The role of the parent FPITT would be to integrate the efforts of these subgroups and to provide an overall perspective of the issues associated with providing the required energy in a variety of scenarios that meet future personal transportation needs.

The hydrogen production program embodies hydrogen generation from a wide range of energy sources including natural gas, coal, biological systems, nuclear heat, wind, solar heat, and grid-based electricity; grid-based electricity employs several of these sources to varying extents, depending on geographical area. In the short term, when a hydrogen pipeline system is not in place, distributed generation in relatively small plants will be required to supplement hydrogen available from existing, large-scale commercial plants. As the fleet of fuel-cell-powered cars grows and hydrogen demand increases, centralized hydrogen-generation plants with pipeline distribution will become increasingly attractive, and these are expected to replace most distributed generation eventually.

Approaches to hydrogen generation using thermal processes include coal and biomass gasification, bio-derived fuels reforming, and thermochemical splitting of water. The DOE had a program, completed in 2009, to improve natural gas reforming. This program established the feasibility of distributed generation at fueling stations using reforming and directionally improved gas cleanup technologies for centralized plants. Commercial options now exist to generate hydrogen either in distributed or centralized plants using natural gas.

The production of hydrogen from coal or from biomass feedstocks appears in the Hydrogen Production Roadmap as both a midterm technology (coal gasification with carbon sequestration) and a long-term technology (biomass gasification with carbon sequestration). The most critical challenges to the use of either feedstock are (1) the capital cost of the gasification processes and (2) the cost and availability of carbon sequestration.

Whereas distributed natural gas reforming has demonstrated the ability to meet the hydrogen cost targets of $2.00-$3.00 per gallon gasoline equivalent (gge) (based on the DOE standard set of assumptions), distributed ethanol reforming has not. The current cost estimates are higher than $4.00/gge, and the targets for 2014 and 2019 are $3.80/gge and $3.00/gge, respectively.

The DOE recognizes that water electrolysis may play an important role in the hydrogen infrastructure and is supporting numerous electrolysis efforts related to capital, electrocatalytic, and configuration/engineering. Some of the challenges with wind and solar-driven electrolysis approaches include efficient power electronics for direct current (dc)-to-dc and alternating current (ac)-to-dc conversion; and controllers and communications protocols to match the source to the electrolyzer.

A significant factor in fuel cost and source-to-wheels efficiency for fuel-cell-powered vehicles is the means for delivering, storing, and dispensing hydrogen. In a fully developed hydrogen economy, the postproduction part of the supply system for high-pressure hydrogen will probably cost as much and consume as much energy as production does (NRC/NAE, 2004).

Progress has been made in all areas of the program. Delivery models have been developed that predict delivery and dispensing costs for different methods as a function of market penetration. In addition, hydrogen compression has been directionally advanced by investigating a centrifugal compressor design and also electrochemical compression.

Past funding for delivery and dispensing apparently has not been based on program needs but on budget constraints. Reducing the cost of delivery and dispensing from the current $2.00-$3.00/kg hydrogen to the 2017 target of less than $1.00/kg hydrogen will require substantial and consistent funding based on program needs. Otherwise, any chance of meeting the 2017 target will be forgone.

Recommendation 4-3. The Fuel Cell Technologies program should adjust its Technology Roadmap to account for the possibility that CO_2 sequestration will not enable a midterm readiness for commercial hydrogen production from coal. It should also consider the consequences to the program of apparent large increases in U.S. natural gas reserves.

Recommendation 4-4. The EERE should continue to work closely with the Office of Fossil Energy to vigorously pursue advanced chemical and biological concepts for carbon disposal as a hedge against the inability of geological storage to deliver a publicly acceptable and cost-effective solution in a timely manner. The committee also notes that some of the technologies now being investigated might offer benefits in the small-scale capture and sequestration of carbon from distributed sources.

Recommendation 4-13. Hydrogen delivery, storage, and dispensing should be based on the program needed to achieve the cost goal for 2017. If it is not feasible to achieve that cost goal, emphasis should be placed on those areas that would most directly impact the 2015 decision regarding commercialization. In the view of the committee, pipeline, liquefaction, and compression programs are likely to have the greatest impact in the 2015 time frame. The cost target should be revised to be consistent with the program that is carried out.

Biofuels and the Partnership

Within the DOE, the Biomass Program has the responsibility for managing the development and progress for the bulk of the needs for biofuels, including biomass production, feedstock logistics, and biomass conversion to a biofuel. Historically the DOE focused on biofuel distribution and end use through the Partnership. This split of focus puts responsibility for making biofuels with the Biomass Program and the responsibility for delivering the biofuel and the light-duty-vehicle drive train with the Partnership.

A thorough systems analysis of the biofuel distribution and end-use system that accounts for engine technologies and petroleum blending fuel properties could help to identify priority areas for further development. This could result in modified priorities for different biomass sources, conversion processes, biofuels, distribution systems, and engines.

Recommendation 4-14. A thorough systems analysis of the complete biofuel distribution and end-use system should be done. This should include (1) an analysis of the fuel- and engine-efficiency gains possible through ICE technology development with likely particular biofuels or mixtures of biofuels and conventional petroleum fuels, and (2) a thorough analysis of the biofuel distribution system needed to deliver these possible fuels or mixtures to the end-use application.

REFERENCES

DOE (U.S. Department of Energy). 2004. *FreedomCAR and Vehicle Technologies Multi-Year Program Plan*. Washington, D.C.: U.S. Department of Energy, Energy Efficiency and Renewable Energy. Available on the Web at <http://www.eere.energy.gov/vehiclesandfuels/resources/fcvt_mypp.shtml>.

NRC (National Research Council). 2005. *Review of the Research Program of the FreedomCAR and Fuel Partnership, First Report*. Washington, D.C.: The National Academies Press.

NRC. 2008. *Review of the Research Program of the FreedomCAR and Fuel Partnership, Second Report*. Washington, D.C.: The National Academies Press.

NRC/NAE (National Research Council/National Academy of Engineering). 2004. *The Hydrogen Economy: Opportunities, Costs, Barriers, and R&D Needs*. Washington, D.C.: The National Academies Press.

1

Introduction

BACKGROUND

The rate at which the world's economies are becoming globalized as well as the industrialization of countries including China and India are bringing an increased demand for energy. Although the forecasting of supply and demand for energy is uncertain, projections of growth in energy use for the next 30 years suggest that for the United States, as well as the rest of the world, it will be a challenge to supply the energy demanded by these economies (NPC, 2007). All sectors of the economy will be affected. The U.S. transportation sector accounted for approximately 28 percent of total U.S. energy use and approximately 71 percent of U.S. petroleum consumption in 2008 (EIA, 2009a). In addition, net U.S. imports of petroleum and refined products have remained high and have accounted for about 56 percent of U.S. petroleum consumption from 2007 to 2009, while U.S. domestic crude oil production has declined. Petroleum prices have also exhibited substantial volatility in the past few years. For example, the preliminary estimate of the crude oil refiners' average acquisition cost in 2008 was $94.74/bbl, a 39 percent increase over the 2007 cost of $67.94/bbl (EIA, 2009a, Table 5.21).[1] Oil and petroleum-derived fuel prices tended to decline in late 2008 and 2009 because of reduced demand caused by the worldwide recession, resulting in an average refiners' acquisition cost of $59.27/bbl in 2009. Diversifying the energy carriers used in mobility systems beyond petroleum-based products and develop-

[1] The crude oil refiners' acquisition cost is the cost to refiners, including transportation and other fees. The average cost is calculated from the sum of the total purchasing (acquisition) costs of all refiners divided by the total volume of all refiners' purchases.

ing new sources for them can be important components of the U.S. energy options and clearly are important national issues.

In addition to energy considerations, the U.S. transportation sector accounts for about 28 percent of total U.S. emissions of carbon dioxide (CO_2), an important greenhouse gas. Concerns about climate change and reducing greenhouse gas emissions have been receiving extensive attention from the administration, Congress, the states, and the Intergovernmental Panel on Climate Change as well as a number of other countries. Legislation has been under consideration by Congress for dealing with CO_2 and other greenhouse gas emissions. Developing vehicles with improved fuel economy that use petroleum-derived gasoline or diesel fuel, or that can use non-petroleum-based energy (e.g., hydrogen fuel, biomass-based fuels, or electricity), have the potential to reduce greenhouse gas emissions from the transportation sector in addition to reducing the nation's dependence on petroleum (NAS/NAE/NRC, 2009a,b; NRC, 2009a).

As President Bush said in his 2001 State of the Union address, hydrogen, as an energy carrier, would have many advantages if it could be developed for the mobility market. However, the challenges of doing so are great. The FreedomCAR and Fuel Partnership was established to address these challenges and to advance the technology enough so that a decision on the commercial viability of hydrogen vehicles can be made by 2015. This report reviews the status and progress of this Partnership. In addition, as discussed in this report, increasing attention under President Obama's administration is being directed toward the use of electricity to power light-duty vehicles with emphasis on plug-in hybrid electric vehicles (PHEVs) and all-electric vehicles (or battery electric vehicles [BEVs]).

The U.S. Department of Energy (DOE) has been involved for more than 30 years in research and development (R&D) programs related to advanced vehicular technologies and alternative transportation fuels. During the 1990s, much of this R&D was conducted under the Partnership for a New Generation of Vehicles (PNGV) program. This initial peacetime government/auto industry partnership was formed between the federal government and the auto industry's U.S. Council for Automotive Research (USCAR).[2] Building on the PNGV program, in January 2002, the Secretary of Energy and executives of DaimlerChrysler, Ford,

[2] USCAR, which predated PNGV, was established by Chrysler Corporation, Ford Motor Company, and General Motors Corporation. Its purpose was to support intercompany, precompetitive cooperation so as to reduce the cost of redundant R&D, especially in areas mandated by government regulation, and to make the U.S. industry more competitive with foreign companies. Chrysler Corporation merged with Daimler Benz in 1998 to form DaimlerChrysler. In 2007, DaimlerChrysler divested itself of a major interest in the Chrysler Group, and Chrysler LLC was formed, which is now Chrysler Group LLC.

The PNGV sought to improve the nation's competitiveness significantly in the manufacture of future generations of vehicles, to implement commercially viable innovations emanating from ongoing research on conventional vehicles, and to develop vehicles that achieve up to three times the fuel efficiency of comparable 1994 family sedans (NRC, 2001; PNGV, 1995; The White House, 1993).

and General Motors announced a new government-industry partnership between DOE and USCAR called FreedomCAR, with "CAR" standing for "Cooperative Automotive Research." In September 2003, FreedomCAR was expanded to also include five large energy companies—BP America, Chevron Corporation, ConocoPhillips, ExxonMobil Corporation, and Shell Hydrogen (U.S.)—to address issues related to supporting the fuel infrastructure. With the recent changes in the automotive industry, USCAR is now USCAR LLC, and the three automotive companies are Chrysler Group LLC, Ford Motor Company, and General Motors Company. The expanded partnership is called the FreedomCAR and Fuel Partnership.[3] With the recent increased interest in technologies for PHEVs and BEVs, which are designed to be plugged into the electric grid to charge up an onboard battery, the electric power sector has become of interest to the Partnership. As a result, a Utility Operations Group has been formed, and two electric utility companies, DTE Energy (Detroit) and Southern California Edison, have joined the Partnership (see Figure 1-1). The long-term vision of the Partnership is a clean and sustainable energy future, in the near term supporting a wide range of hybrid electric vehicles and with a long-term strategic goal of developing technologies for hydrogen-powered fuel cell vehicles that are not dependent on oil and with no harmful emissions or greenhouse gases (DOE, 2004a,b,e). Furthermore, "the aim is to achieve this technology shift without sacrificing mobility or freedom of choice for American consumers" (DOE, 2004e, p. 18).

The Partnership addresses the development of advanced technologies for all light-duty passenger vehicles: cars, sport utility vehicles (SUVs), pickups, and minivans. It also addresses technologies for hydrogen production, distribution, dispensing, and storage, and now, with the interest in electric vehicles (primarily since the change in administrations), it is addressing the interface and infrastructure issues associated with the electric utility industry.

The Partnership started with a presidential commitment to request $1.7 billion over 5 years (FY 2004 to FY 2008), with appropriations thus far of about $243 million, $307 million, and $339 million for FY 2004, FY 2005, and FY 2006, respectively. Funding for FY 2007 was about $401 million, FY 2008 was about $419 million, and FY 2009 funding was about $474 million; appropriations for FY 2010 are about $467 million (see Chapter 5). In addition, although under the American Recovery and Reinvestment Act of 2009 (ARRA, or the "Recovery Act"; Public Law 111-5) funds have not been appropriated directly to the Partnership, $2.8 billion of funding has been provided to related activities including

---

[3] In February 2003, before the announcement of the FreedomCAR and Fuel Partnership, President Bush announced the FreedomCAR and Hydrogen Fuel Initiative to develop technologies for (1) fuel-efficient motor vehicles and light trucks, (2) cleaner fuels, (3) improved energy efficiency, and (4) hydrogen production and a nationwide distribution infrastructure for vehicle and stationary power plants, to fuel both hydrogen internal combustion engines and fuel cells (DOE, 2004a). The expansion of the FreedomCAR and Fuel Partnership to include the energy sector after the announcement of the initiative also supports the goal of the FreedomCAR and Hydrogen Fuel Initiative.

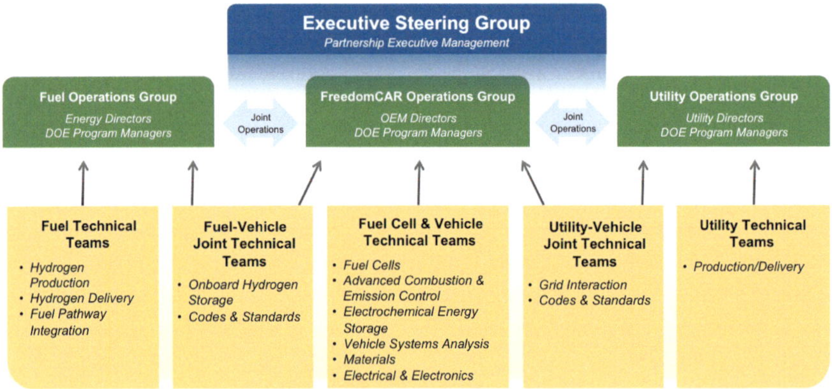

FIGURE 1-1 The organizational structure of the FreedomCAR and Fuel Partnership. NOTE: OEM, original equipment manufacturer. SOURCE: P. Davis and S. Satyapal, DOE, "Overview of the FreedomCAR and Fuel Partnership," Presentation to the committee, August 4, 2009, Southfield, Michigan.

automotive battery manufacturing facilities and transportation electrification.[4] Funding for research, development, and demonstration activities goes to universities, the national laboratories, and private companies. Especially in the case of development activities, projects are often cost-shared between the private sector and the federal government (see Chapter 5 for further discussion).

The Partnership plays an important role in the planning, pursuit, and assessment of high-risk, precompetitive R&D for many of the needed vehicle and fuel technologies. Federal funds enable this work to move forward. However, with the change in administrations along with the economic problems of the automotive original equipment manufacturers (OEMs), it appears that much of the emphasis could shift to PHEV and BEV technologies, which are apparently viewed by the new administration as nearer-term. The Partnership also serves as a communication mechanism for those interested, including government, the private sector, the national laboratories, universities, the public, and others. In addition, the success of the FreedomCAR and Fuel Partnership can serve as an inspiration and motivation for the next generation of scientists and engineers, and thus contribute to restoring American leadership in research and its application for the public good.

In late 2008 the National Research Council (NRC) formed the Committee on Review of the FreedomCAR and Fuel Research Program, Phase 3 (see Appendix A for biographical information on the members). Its report represents the third

[4] S. Satyapal and P. Davis, "Review of the FreedomCAR and Fuel Partnership, Phase 3," Presentation to the committee, April 27, 2009, Washington, D.C.

review by the NRC of the research program of the Partnership. The main charge to the committee for the Phase 3 report is to review activities between Phases 2 and 3, which included very little related to PHEV, BEV, or biofuel technologies. The first review was conducted during 2004-2005 and the second review during 2007-2008, resulting in the Phase 1 and Phase 2 reports (NRC, 2005, 2008a). (The first review will be referred to as the Phase 1 review or report and the second review as the Phase 2 review or report.)

COMMITTEE'S INTERIM LETTER REPORT

Unlike the experience with the Phase 1 and Phase 2 reviews, this committee was asked to provide an interim letter report that broadly reviewed the strategy and structure of the Partnership before undertaking the full in-depth technical review of the Partnership. The committee held a meeting on April 27-28, 2009, to hear presentations from the Partnership and to work on its letter report, which was addressed to Steven Chu, Secretary of Energy, and Cathy Zoi, Assistant Secretary of Energy Efficiency and Renewable Energy. The letter report, issued on July 10, 2009, is contained in Appendix B.

The committee's major messages in its interim report (NRC, 2009b, pp. 1-2) are as follows:

> The committee recognizes and agrees with the new Administration's focus on nearer-term technologies. However, it also emphasizes the need for continued investment in longer-term, higher-risk, higher-payoff vehicle technologies that could be highly transformational with regard to reduced use of petroleum and reduced emissions. Such technologies include advanced batteries, technologies for hydrogen storage, and hydrogen/fuel cells. The committee has also concluded that for researchers, contractors, and investors to be willing to make long-term commitments to these and other potentially important developing technologies, a consistent year-to-year level of support must be provided.
>
> The committee has further concluded that, given increasing concerns about greenhouse gas (GHG) emissions and world climate change, the Partnership should incorporate in its planning a broader-scope, "cradle-to-grave" analysis rather than a "well-to-wheels" approach, to better consider total emissions and the full environmental impact of using various fuels and technologies. In addition, the Partnership should consider broadening the scope of technical approaches being considered within each of what the committee considers to be the three major fuel and vehicle pathways—biofuels/internal combustion engine (ICE) vehicles, plug-in electric vehicles (PHEVs)/battery electric vehicles (BEVs), and hydrogen-fueled fuel cell vehicles.
>
> Finally, the committee concluded that several measures should be considered by DOE to assist in implementing these suggestions. One is to provide temporary reductions in cost-share requirements to ease the burden on prospective researchers. Otherwise, there could be a significant number of potential worthy contributors who cannot afford the matching funds. Another implementation suggestion, occasioned by the obvious financial problems of the automotive companies (OEMs), is to consider providing direct funding to them to help keep important in-house research programs active. Other suggestions are included in the balance of the report.

GOALS AND TARGETS

The long-term goal of the Partnership is to enable the transition to light-duty passenger vehicles that operate free of petroleum and free of harmful emissions (DOE, 2004b). Taking steps to begin to reduce the nation's dependence on imported petroleum is central to this goal. The current plan envisions a pathway starting with more fuel-efficient internal combustion engines (ICEs) and hybrid electric vehicles (HEVs), including PHEVs, potential use of all-electric-drive vehicles, the deployment of biofueled ICE vehicles, and, ultimately, the addition of an infrastructure for supplying hydrogen fuel for fuel-cell-powered vehicles (DOE, 2004b).[5] Although not part of the original FreedomCAR and Fuel Partnership charter, the existence of an adequate electrical infrastructure to provide recharging energy for PHEVs and BEVs is clearly essential for the Partnership goals. To this end, the Partnership works with other DOE offices and also sponsors some research to ensure that such infrastructure is in place when needed, or to learn what it will take to ensure that it can be in place when needed. If biofuels are to supply a significant portion of the U.S. transportation fuel needs, the infrastructure for the harvesting of biomass, its conversion, and its wide-scale distribution, probably by pipelines, will have to be put in place (NAS/NAE/NRC, 2009b; NRC, 2008b). Thus hydrogen-fueled fuel cell vehicles, plug-in or all-electric vehicles, and biofueled vehicles all will have to face infrastructure issues and hurdles to varying degrees. Heretofore, the infrastructure issues associated with PHEVs, BEVs, and biofuels have not been part of the FreedomCAR and Partnership charter, but those issues are essential to meeting Partnership goals.

To address the technical challenges associated with this envisioned pathway, the Partnership has established quantitative technology and cost targets[6] for 2010 and 2015 in eight areas:

- Fundamental combustion and emission control R&D for ICEs,
- Fuel cell power systems,
- Fuel cells,
- Hydrogen storage systems,
- Energy storage systems for hybrid vehicles,
- Hydrogen production and delivery systems,
- Electric propulsion systems, and
- Materials for lightweight vehicles.

---

[5] J. Sakioka, R. Modlin, and B. Peirce, "Automotive OEM Perspective on FreedomCAR and Fuels Program," Presentation to the committee, August 4, 2009, Southfield, Michigan.

[6] All references to cost imply estimated variable cost (or investment, as appropriate) based on high volume (500,000 annual volume) unless otherwise stated. "Cost" refers to the cost of producing an item, whereas "price" refers to what the consumer would pay.

These goals and the research related to their attainment are discussed later in this report. Given some of the changes in focus of the Partnership, some goals and targets for individual technologies are being reevaluated by the Partnership. Technical teams, as noted in the next section, "Organization of the Partnership," specify and manage technical and crosscutting needs of the program.

ORGANIZATION OF THE PARTNERSHIP

The Partnership consists of a number of oversight groups and technical teams that have participants from government and industry (see Figure 1-1). The Executive Steering Group, which is responsible for the governance of the Partnership, is made up of the DOE assistant secretary for the Office of Energy Efficiency and Renewable Energy (EERE) and a vice-presidential-level executive from each of the Partnership companies. The FreedomCAR Operations Group, made up of DOE program managers and directors from USCAR member companies, is responsible for directing the technical teams and prioritizing research issues. The Fuel Operations Group, made up of DOE program managers and energy company directors, is responsible for the direction of the fuel technical teams. And recently, in February 2009, the Utility Operations Group was added to the organization with two utility companies, DTE Energy (Detroit) and Southern California Edison, to address the coordination between electric-based vehicle technology and the electric utility infrastructure (DOE, 2009). In the past, the FreedomCAR Operations Group and the Fuel Operations Group have periodically held joint meetings to coordinate fuel and power plant issues and to identify strategic or policy issues that warrant attention by the Executive Steering Group (DOE, 2004c). With the addition of the Utility Operations Group, the committee expects that joint meetings among all of the operations groups will take place.

The Partnership has formed industry-government technical teams responsible for setting technical and cost targets as well as focusing appropriate R&D on the candidate subsystems (see Figure 1-1). Most of these technical teams focus on specific technical areas, but some, such as codes and standards and vehicle systems analysis, focus on crosscutting issues. A technical team consists of scientists and engineers with technology-specific expertise from the USCAR member companies, energy partner companies, utility industry companies, and national laboratories, as well as DOE technology development managers. Team members may come from other federal agencies if approved by the appropriate operations group(s). A technical team is responsible for developing R&D plans and roadmaps, reviewing research results, and evaluating technical progress toward meeting established research goals (DOE, 2004c). Its discussions are restricted to nonproprietary topics.

Members of the fuel cell and vehicle technical team come from the USCAR partners and the DOE. They handle fuel cells, advanced combustion and emissions control, systems engineering and analysis, electrochemical energy storage, materials, and electrical systems and power electronics. The three fuel technical teams address hydrogen production, hydrogen delivery, and fuel/vehicle pathway integration, each of which has members from the energy companies and the DOE. There are two joint technical teams connecting the fuel teams and the vehicle teams: an onboard hydrogen storage team and a codes and standards team. The utility interface issues have resulted in new technical teams related to electricity, namely, grid interaction, codes and standards, and production/delivery.

At the DOE, primary responsibility for the FreedomCAR and Fuel Partnership rests with the EERE.[7] The two main program offices within EERE that manage the Partnership are the Vehicle Technologies (VT) program and the Hydrogen, Fuel Cells, and Infrastructure Technologies program (HFCIT; this is now called the Fuel Cell Technologies [FCT] program).

The VT program has the following specific goal: to support "R&D that will lead to new technologies that reduce our nation's dependence on imported oil, further decrease vehicle emissions, and serve as a bridge from today's conventional power trains and fuels to tomorrow's hydrogen-powered hybrid fuel cell vehicles" (DOE, 2004b, p. ES-2). The VT also includes the 21st Century Truck Partnership.[8]

The FreedomCAR and Fuel Partnership activities in the VT program are organized into these areas:

- Vehicle systems analysis and testing to provide an overarching vehicle systems perspective to the technology R&D subprograms and other activities in the VT and FCT programs;
- Advanced energy-efficient, clean ICE power trains using various petroleum and non-petroleum-based fuels, including hydrogen and/or electricity;
- Electrochemical energy storage technologies (batteries and ultracapacitors);
- Advanced power electronics and electric machines;
- Materials technology for lightweight vehicle structures and for propulsion system components, including power electronics and ICEs; and
- Fuel technologies that enable current and emerging advanced ICEs and emission control systems to be as efficient as possible while meeting future emission standards and that reduce reliance on petroleum-based fuels.

---

[7] The EERE has a wide variety of technology R&D programs and activities related to renewable energy technologies, ranging from the production of electricity from solar energy or wind and the production of fuels from biomass, to the development of technology to enhance energy efficiency, whether for vehicles, appliances, buildings, or industrial processes. It also has programs on distributed energy systems (see Appendix C for an EERE organizational chart).

[8] The DOE supports several other programs related to the goal of reducing dependence on imported oil. The 21st Century Truck Partnership supports R&D on more-efficient and lower-emission commercial road vehicles. The NRC Committee to Review the 21st Century Truck Partnership has reviewed that program (NRC, 2008c).

The FCT program directs activities in hydrogen production, storage, and delivery and integrates these efforts with transportation and fuel cell development activities. The proton exchange membrane (PEM) fuel cell R&D is undertaken in the FCT program, which is focused on the following:

- Overcoming technical barriers through R&D on hydrogen production, delivery, and storage technologies, as well as on fuel cell technologies for transportation, distributed stationary power, and portable power applications;
- Addressing safety concerns and developing model codes and standards;
- Validating and demonstrating hydrogen fuel cells in real-world conditions; and
- Educating key stakeholders whose acceptance of these technologies is critical to their success in the marketplace (DOE, 2004a,b).

The manager of FCT is the overall DOE hydrogen technology program manager.

Some activities related to the FCT program focus are not within the EERE. The Office of Fossil Energy (FE) supports the development of technologies to produce hydrogen from coal and to capture and sequester carbon. The Office of Nuclear Energy (NE) supports research into the potential use of high-temperature nuclear reactors to produce hydrogen, while the Office of Science (SC) supports fundamental work on new materials to store hydrogen, catalysts, fundamental biological or molecular processes for hydrogen production, fuel cell membranes, and other related basic science areas (DOE, 2004d,e). Within the EERE there also is an Office of Biomass Energy, which is not part of the FreedomCAR and Fuel Partnership. However, biomass is of interest to the Partnership, both as one possible source of hydrogen as well as of biomass-based liquid transportation fuels (e.g., ethanol) and as part of a strategy to diversify energy sources for the transportation sector; thus there is cooperation between the Partnership and the biomass program. The committee believes, as discussed in the report, that improving ICE vehicles using biomass-based fuels is an important part of the portfolio of vehicle technologies that need to be addressed, as mentioned in the committee's interim letter report (see Appendix B; NRC, 2009b). And now with the importance of understanding the interface between electric vehicle technology and the electric utility sector, DOE's Office of Electricity Delivery and Reliability, whose focus is on the U.S. electric transmission and distribution system, is another office that needs to interface with the Partnership's efforts. This office is a separate office, as is the EERE, within the Office of the Undersecretary of Energy.

RECENT INITIATIVES

External developments that may affect the Partnership program have continued to emerge since the publication of the Phase 1 and 2 reviews by the NRC

(2005, 2008a). Some of these are enumerated in the committee's interim letter report (NRC, 2009b; see Appendix B). For many years there has been concern, now growing rapidly, on the part of both the Congress and the administration with regard to the security implications of U.S. dependence on imported energy, especially petroleum. Adding to these concerns are the issues of the emissions of greenhouse gases and the apparent effects on global warming. Increases of about 40 percent by 2016 in the corporate average fuel economy (CAFE) standards for light-duty vehicles are being implemented, and Congress has supported legislation that requires increasing the production of fuels from renewable, bio-based sources and other alternative fuels as part of this effort to reduce petroleum-based gasoline consumption.

Congress has supported the expanded production of fuel ethanol, which increased rapidly during the past few years and reached about 9 billion gal/yr in 2008, and is providing incentives for much more expansion.[9] Although ethanol production in the United States is now mostly from corn, eventually ethanol is expected to be produced from cellulose (e.g., grasses, woody plants, and agricultural and wood wastes). Such processes are not yet developed and will require substantial R&D to be successful. Other potential alternative fuels include gasoline or diesel liquids derived from coal or oil shale. Many alternatives are being explored, but which fuels and to what extent and at what cost they will be able to enter the marketplace over the coming decades remain very uncertain (NAS/NAE/NRC, 2009a,b).

In addition, there are numerous bills in Congress aimed at achieving significant reductions in greenhouse gas emissions. If passed, these bills will create incentives either to improve the fuel economy of vehicles or to stimulate the adoption of fuels that produce less greenhouse gases than those from gasoline and diesel fuel. The Environmental Protection Agency has also announced that it plans to regulate greenhouse gas emissions.

There has also been increasing interest in PHEVs, which would contain an ICE and a battery that could be charged from the electric grid when not in use. Depending on the battery capacity and control logic, a version of this car could be driven between 10 and 40 miles on battery power alone, which covers the distance that most people drive to work every day and much of all daily travel in the United States. A cost-effective, durable battery of adequate capacity would enable the electric grid to supply a significant part of the energy for U.S. vehicles. Since virtually no petroleum is used to produce electricity in the United States, this would reduce demand for petroleum in the transportation sector but would not necessarily decrease the amount of CO_2 production. During the Phase 2 review, the committee noted that, depending on the mix of fuels used to supply electricity for such vehicles, this could lead to increased natural gas imports and consumption of coal, with implications for greenhouse gas emissions. However, recent forecasts

[9] See the Renewable Fuels Association Web site at <http:www.ethanolrfa.org/industry/statistics/#D>.

by the Energy Information Administration (EIA) on a better outlook for domestic natural gas production leads the EIA to forecast a decline in U.S. imports for natural gas over the next two decades (EIA, 2009b). The extent to which penetration of PHEVs into the marketplace would affect U.S consumption and imports of natural gas is uncertain at this time. The Energy Policy Act of 2005 (Public Law 109-58) called for a research program on such vehicles as well as flexible-fuel vehicles (e.g., vehicles that can use gasoline or ethanol or a mixture of both). The Phase 1 and Phase 2 reviews of the Partnership also called for increased research on such high-energy storage batteries. The Obama administration has also stressed PHEVs as what it views as a lower-risk, nearer-term technology and has as a goal to have 1 million PHEVs on the road by 2015.[10] In fact, as discussed in the committee's interim report, the administration's focus on what it considers to be the nearer-term technologies led it to eliminate funding for the Partnership's R&D on hydrogen-fueled fuel cell vehicles in the congressional budget request for the FY 2010 budget. However, Congress appropriated somewhat reduced funds to these technology areas in the FY 2010 appropriations bills, and the administration has requested even further reduced funding for FY 2011.

As discussed in the committee's interim letter report, the turmoil in the automotive industry is also of concern. The bankruptcy of the General Motors Corporation and Chrysler and their restructuring, as well as the sharp decline in demand for new vehicles throughout the U.S. and world economies, have led to a significant constraint on resources, not only for the three U.S. automotive manufacturers but also for automotive suppliers as well as foreign companies. It is not yet clear what the extent of this economic downturn will be in the automotive sector with regard to constraining investments in high-risk, long-term automotive technologies, including activities within the Partnership.

This increased interest on the part of the public, Congress, and the administration in reducing petroleum use, and hence energy imports and greenhouse gas emissions, could further stimulate interest in the development of hydrogen-fueled vehicles. But it will likely also stimulate interest in biofuels, alternative liquid fuels, PHEVs, and all-electric vehicles, thus creating a funding competition for hydrogen-fueled fuel cell vehicles. As noted in recent NRC reports and in the committee's interim letter report, a balanced portfolio of R&D on a variety of long-range options will be needed (NRC, 2008b; NAS/NAE/NRC, 2009a,b; NRC, 2009a,b). In addition, it is likely that different vehicle technologies will have different competitive advantages in different market segments. For example, all-electric vehicles may find a more suitable market in intra-urban transportation where a limited vehicle driving range may be acceptable to the automotive purchaser.

---

[10] See, for example, "Obama-Biden New Energy for America" on the Web at <http://www.barackobama.com/pdf/factsheet_energy_speech_080308.pdf>. Predicting whether 1 million PHEVs will be on the road by 2015 is difficult since it depends on many uncertainties including technical performance, cost, consumer behavior, subsidies, policies, and other factors.

VEHICLES AND FUELS

The Phase 1 review of the Partnership contains some general discussion of the importance of linking vehicles, fuels, and infrastructure to ensure that the impacts on the commercial market will be significant and widespread. (That discussion is not repeated here; the reader is referred to the Phase 1 report for that background [NRC, 2005, Chapter 1].) Successful examples of new fuels include the introduction of unleaded gasoline in 1971 and the introduction of reformulated gasoline in the 1990s. But efforts to introduce alternative fuels such as methanol, ethanol, and compressed natural gas (CNG) on a wide scale, with the exception of small percentages of ethanol as a gasoline additive, have all foundered. Alcohol fuels, such as 85 percent methanol (M85) or 85 percent ethanol (E85), work well in vehicles designed to accept them, and although there are several million vehicles on the road that can use these fuels, no extensive fueling infrastructure has developed. In spite of its clean-burning properties and its relatively low unit energy costs, CNG vehicles have also enjoyed limited success. They are mainly found in fleets and in niche markets. This need for both the acceptance of new vehicle technology that relies on nontraditional fuel and the widespread availability of that fuel in the marketplace is the reason that the Partnership supports R&D for both vehicles and fuels and, now, for the interface with the electric utility industry. The program seeks ultimately to enable the widespread deployment of a number of different vehicle options. The primary long-term focus up until recently was on fuel cell vehicles fueled by convenient, competitively priced hydrogen. The Partnership is structured to address the obvious barriers to achieving this goal for both the fuel cell vehicle and the hydrogen fuel production and delivery systems. Other alternative fuels, such as cellulosic-based ethanol, also will require extensive infrastructure investments if they become a significant part of the light-duty-vehicle fuel supply (NAS/NAE/NRC, 2009a).

There is now more focus on PHEVs, however, especially in the nearer term, since the U.S. infrastructure for electricity supply, transmission, and distribution is already in place. Other alternative fuels, such as cellulosic-based ethanol, also will require extensive infrastructure investments if they become a significant part of the light-duty-vehicle fuel supply (NAS/NAE/NRC, 2009a).

Hydrogen represents a completely new fuel for the transportation sector, and a completely new infrastructure will have to be put in place—creating a chicken-and-egg situation. Even if successful and cost-competitive fuel cell vehicles are developed, they cannot be sold in great numbers if no fuel infrastructure exists. Likewise, an extensive hydrogen fuel infrastructure cannot be economically justified to service the first few fuel-cell-powered vehicles that might be built. *The Hydrogen Economy: Opportunities, Costs, Barriers, and R&D Needs* (NRC/NAE, 2004) and *Transitions to Alternative Transportation Technologies—A Focus on Hydrogen* (NRC, 2008b) emphasized the importance of the distributed production of hydrogen: for example, using natural gas and the existing natural gas infrastructure to produce hydrogen at fueling stations; or using renewable energy—for example, wind to electric systems—to generate hydrogen through electrolysis at the fueling stations using the existing electrical grid infrastructure. Generating hydrogen at the fueling station would avoid the need initially to install a vast hydrogen distribution infrastructure. The DOE has focused significant efforts on this transition concept, as discussed in Chapter 4. Nevertheless, even assuming a maximum practicable number of hydrogen-fueled fuel cell vehicles beginning to enter the marketplace in 2015, it would take a couple of decades for significant impacts on reductions in petroleum consumption and greenhouse gas emissions (NRC, 2008b).

The other major nonpetroleum approach to fueling light-duty vehicles is to produce liquid fuels (e.g., ethanol) from cellulosic biomass, from coal, or from a combination of coal and biomass. The NRC recently completed a study on the various technologies and costs for producing these fuels and the timescale and potential impacts on petroleum consumption and greenhouse gas emissions (NAS/NAE/NRC, 2009a). These studies consider the full fuel cycle from source to wheels to calculate emissions and cost to drive a vehicle a mile. Such analyses must take into account the volumetric energy density of the fuel since, for example, a gallon of ethanol has about two-thirds of the energy of a gallon of gasoline. Thus, a car, all other things being equal, could drive a greater distance on a gallon of gasoline compared to a gallon of ethanol. Such factors are taken into account in ongoing analyses by the Partnership on the full fuel cycle analyses of energy, CO_2 and other emissions, and costs for light-duty vehicles (DOE, 2004b). Even assuming that technical and cost barriers were overcome, such approaches to fueling the transportation sector would take two to three decades to make a significant impact.

With regard to PHEVs, another NRC committee completed a recent study that investigated the potential costs and impacts on U.S. greenhouse gas emissions and petroleum consumption from 2010 to 2050 (NRC, 2009a). The study shows that PHEV-40 vehicles are likely to be quite costly initially, at about $18,000 more than an equivalent conventional vehicle, although a PHEV-10 will have a much more modest cost increment of about $6,300.[11] There will also be required some modest electrical system upgrades for some homes, and millions of light-duty-vehicle owners do not live in houses, or houses with garages. The scenarios in the NRC (2009a) study indicate that PHEV-40s are unlikely to achieve cost-effectiveness before 2040 at gasoline prices below $4.00/gal, but PHEV-10s can achieve it before 2030. Thus, it would be several decades before lifetime fuel savings started to balance the higher first cost of the vehicles, and subsidies of tens to hundreds of billions of dollars over several decades would be needed for the transition. Another conclusion of that study is that PHEVs will have little impact on oil consumption before 2030, although more substantial reductions could be achieved by 2050.

[11] A PHEV-10 has a battery that can provide an all-electric driving range of 10 miles; a PHEV-40 has an all-electric driving range of 40 miles.

In addition, although PHEV-40s are more effective than PHEV-10s compared to conventional vehicles with regard to the emissions on a total "source-to-wheels" basis, the greenhouse gas benefits are small for a couple of decades unless the electrical grid is decarbonized with renewable energy, nuclear plants, or fossil-fuel-fired power plants with carbon capture and storage (CCS, also referred to as carbon capture and sequestration) systems.

Consequently, by far the greatest contribution to reduced energy use (especially that of petroleum) and emissions reductions by and from the U.S. vehicle fleet over the next 20 years and beyond will come from continued improvement in ICEs, hybrid electric vehicles, and their fuels. To reduce transportation fuel use, current industry-wide efforts to improve the efficiency of ICEs and to develop the corresponding fuels further must continue or, even better, accelerate. This is true regardless of the degree to which HEV power trains proliferate or whether advanced diesel engines achieve customer acceptance and meet emissions standards. The urgency of this task is amplified by the reality that, with the current reduced new-vehicle sales of about 10 million in the United States every year, it would take about 20 years to turn over the national fleet of roughly 225 million light-duty vehicles. If the U.S. marketplace recovers to new-vehicle sales of about 16 million per year as it was before the current worldwide recession, the turnover time would be about 15 years.

While much of the Partnership activity is devoted to fuel cell vehicles and hydrogen fuel, advanced vehicles such as PHEVs and BEVs, and biofueled vehicles, further improvement in conventional ICEs and HEVs could contribute significantly to the goals of energy independence and reduced carbon emissions and should benefit from the continued collaboration between industry engineers and the DOE national laboratories in this area. The status of Partnership efforts to develop ICEs and emission control technologies is discussed in Chapter 3.

The goal toward low emissions, whether of CO_2 or various air and water pollutants that arise as a result of the full fuel cycle and life cycle of vehicles and their fuels, will require fundamental changes in the manner in which vehicle fuels (or electricity) are produced. If a transition to hydrogen-fueled fuel cell vehicles is to result in low emissions for the full fuel cycle, then hydrogen will have to be produced with processes having low emissions—for example, in central plants fueled by coal or natural gas with CCS, or by using renewable energy or nuclear energy technologies (NRC, 2008b). For PHEVs or BEVs, the manner in which electricity is produced will determine to what extent such vehicles will reduce carbon emissions. Thus, the electric power system will have to transition to much greater use of low-carbon systems, such as fossil fuel plants with CCS, renewable energy, or nuclear energy (NRC, 2009a). The same argument holds true for biofuels, which have the advantage that the biomass crops absorb CO_2 from the atmosphere, and thus fuels derived from biomass have the potential to have a lower carbon footprint (NAS/NAE/NRC, 2009a). Other liquid fuels, for example those produced from coal, could have CO_2 emissions equivalent to petroleum if CCS is used, or, in the case of mixtures of coal/biomass conversion plants with CCS, fuel could have substantially lower CO_2 emissions than those from petroleum-based fuels (NAS/NAE/NRC, 2009a). Thus, no matter which advanced vehicles are considered, the production of either the fuel or the electricity to supply the vehicles will have to be substantially changed to meet significant reductions in emissions, especially of carbon.

COMMITTEE APPROACH AND ORGANIZATION OF THIS REPORT

The statement of task for this committee is as follows:

> The National Academies' National Research Council (NRC) Committee on Review of the Research Program of the FreedomCAR and Fuel Partnership, Phase 3, will address the following tasks [Note that the committee's interim letter report issued on July 10, 2009, addressed Item 6 in the statement of task.]:
>
> (1) Review the challenging high-level technical goals and timetables for government and industry R&D efforts, which address such areas as (a) integrated systems analysis; (b) fuel cell power systems; (c) hydrogen storage systems; (d) hydrogen production and distribution technologies necessary for the viability of hydrogen-fueled vehicles; (e) the technical basis for codes and standards; (f) electric propulsion systems; (g) electric energy storage technologies; (h) lightweight materials; and (i) advanced combustion and emission control systems for internal combustion engines (ICEs).
>
> (2) Review and evaluate progress and program directions since the Phase 1 and 2 reviews toward meeting the Partnership's technical goals, and examine ongoing research activities and their relevance to meeting the goals of the Partnership.
>
> (3) Examine and comment on the overall balance and adequacy of the research and development effort, and the rate of progress, in light of the technical objectives and schedules for each of the major technology areas.
>
> (4) Examine and comment, as necessary, on the appropriate role for federal involvement in the various technical areas under development, especially in light of activities ongoing in the private sector or in the states.
>
> (5) Examine and comment on the Partnership's strategy for accomplishing its goals, especially in the context of ongoing developments in biofuels, plug-in hybrid electric vehicles, electric vehicles, the recent enactment of legislation on corporate average fuel economy standards for light-duty vehicles, and possible legislation on carbon emissions. Other issues that the committee might address include (a) program management and organization; (b) the process for setting milestones, research directions, and making Go/No Go decisions; (c) collaborative activities needed to meet the program's goals (e.g., among the various offices and programs in DOE, the U.S. Department of Transportation, USCAR, the fuels industry, electric power sector, universities, other parts of the private sector [such as venture capitalists], and others); and (d) other topics that the committee finds important to comment on related to the success of the program in meeting its technical goals.
>
> (6) As a first step in examining the Partnership's strategy, and given the changes that may take place with the new Administration, the committee at its first full committee meeting will address potential changes in the program strategy and program structure. The committee will write a short interim letter report with suggestions and recommendations on program strategy and structure and aim to deliver it to the sponsor within 1 month after the meeting. The date of delivery of the letter report will be contingent on when

the meeting is scheduled and timely input of information from the representatives of the Partnership.

(7) Review and assess the actions that have been taken in response to recommendations from the NRC Phase 2 review of the Partnership.

(8) Write a final report documenting its conclusions and recommendations.

The committee met three times to hear presentations from DOE and industry representatives involved in the management of the program and to discuss insights gained from the presentations and the written material gathered by the committee. It met a fourth time to review drafts of the report sections (see Appendix E for a list of committee meetings and presentations). The committee also had one meeting in April 2009 before writing its interim report (see Appendix B). The committee established subgroups to investigate specific technical areas and formulate questions for the program leaders to answer. The subgroups also met with the Partnership technical team leaders to clarify answers to questions and better understand the team dynamics, and several committee members visited the General Motors Honeoye facility in New York State to view its fuel cell vehicle developments. The Partnership also provided responses to the recommendations from the Phase 2 report, and these are included in the National Academies' public access file.

The Summary presents the committee's main conclusions and recommendations. This chapter (Chapter 1) provides background on the FreedomCAR and Fuel Partnership, on its organization, and on the dual nature—vehicle development and fuel development—of the program. Chapter 2 examines the important crosscutting issues that the program is facing. Chapter 3 looks more closely at R&D for the various vehicle technologies, and Chapter 4 examines R&D for hydrogen production, distribution, and dispensing, as well as issues related to the use of biofuels in internal combustion engines. Finally, Chapter 5 presents an overall assessment. In addition to the appendixes referred to above (committee biographical information, the interim letter report, the EERE organizational chart, and the list of meetings and presentations), two additional appendixes are included: Appendix D contains the Phase 2 recommendations, and Appendix F defines the report's acronyms and abbreviations.

REFERENCES

DOE (Department of Energy). 2004a. *Hydrogen, Fuel Cells and Infrastructure: Multi-Year Research, Development and Demonstration Plan*. DOE/GO-102003-1741. Washington, D.C.: U.S. Department of Energy, Energy Efficiency and Renewable Energy. Available on the Web at <http://www.eere.energy.gov/hydrogenandfuelcells/mypp/>.

DOE. 2004b. *FreedomCAR and Vehicle Technologies Multi-Year Program Plan*. Washington, D.C.: U.S. Department of Energy, Energy Efficiency and Renewable Energy. Available on the Web at <http://www1.eere.energy.gov/vehiclesandfuels/resources/fcvt_mypp.html>.

DOE. 2004c. *Partnership Plan. FreedomCAR & Fuel Partnership*. Washington, D.C.: U.S. Department of Energy, Energy Efficiency and Renewable Energy. Available on the Web at <http://www.eere.energy.gov/vehiclesandfuels/pdfs/program/fc_fuel_partnership_plan.pdf>.

DOE. 2004d. *Basic Research Needs for the Hydrogen Economy: Report of the Basic Energy Sciences Workshop on Hydrogen Production, Storage, and Use,* May 13-15, 2003. Washington, D.C.: U.S. Department of Energy, Office of Science. Available on the Web at <http://www.er.doe.gov/production/bes/hydrogen.pdf>.

DOE. 2004e. *Hydrogen Posture Plan: An Integrated Research, Development and Demonstration Plan*. Washington, D.C.: U.S. Department of Energy. Available on the Web at <http://www.eere.energy.gov/hydrogenandfuelcells/pdfs/hydrogen_posture_plan.pdf>.

DOE. 2009. "Addendum to the FreedomCAR and Fuel Partnership Plan to Integrate Electric Utility Industry Representatives." February. Washington, D.C.: U.S. Department of Energy. Available on the Web at <http://www1.eere.energy.gov/vehiclesandfuels/pdfs/program/fc_fuel_addendum_2-09.pdf>.

EIA (Energy Information Administration). 2009a. *Annual Energy Review 2008,* Washington, D.C. Available on the Web at <http://tonto.eia.doe.gov/FTPROOT/multifuel/038408.pdf>.

EIA. 2009b. *Annual Energy Outlook for 2009 with Projections to 2030: Updated Annual Energy Outlook 2009 Reference Case with ARRA*. Washington, D.C. Available on the Web at <http://www.eia.doe.gov/oiaf/servicerpt/stimulus/excel/aeostimtab_13.xls>.

NAS/NAE/NRC (National Academy of Sciences/National Academy of Engineering/National Research Council). 2009a. *Liquid Transportation Fuels from Coal and Biomass: Technological Status, Costs, and Environmental Impacts.* Washington, D.C.: The National Academies Press.

NAS/NAE/NRC. 2009b. *America's Energy Future: Technology and Transformation.* Washington, D.C.: The National Academies Press.

NPC (National Petroleum Council). 2007. Facing the Hard Truths About Energy: A Comprehensive View to 2030 of Global Oil and Natural Gas. Executive Summary, July 18. Available on the Web at <http://downloads.connectlive.com/events/npc071807/pdf-downloads/Facing_Hard_Truths-Executive_Summary.pdf>.

NRC (National Research Council). 2001. *Review of the Research Program of the Partnership for a New Generation of Vehicles, Seventh Report*. Washington, D.C.: National Academy Press.

NRC. 2005. *Review of the Research Program of the FreedomCAR and Fuel Partnership, First Report.* Washington, D.C.: The National Academies Press.

NRC. 2008a. *Review of the Research Program of the FreedomCAR and Fuel Partnership, Second Report*. Washington, D.C.: The National Academies Press.

NRC. 2008b. *Transitions to Alternative Transportation Technologies—A Focus on Hydrogen.* Washington, D.C.: The National Academies Press.

NRC. 2008c. *Review of the 21st Century Truck Partnership.* Washington, D.C.: The National Academies Press.

NRC. 2009a. *Transitions to Alternative Transportation Technologies—Plug-in Hybrid Electric Vehicles.* Washington, D.C.: The National Academies Press.

NRC. 2009b. *Letter Report on Review of the Research Program of the FreecomCAR and Fuel Partnership, Phase 3.* Washington, D.C.: The National Academies Press.

NRC/NAE (National Research Council/National Academy of Engineering). 2004. *The Hydrogen Economy: Opportunities, Costs, Barriers, and R&D Needs*. Washington, D.C.: The National Academies Press.

PNGV (Partnership for a New Generation of Vehicles). 1995. *Partnership for a New Generation of Vehicles Program Plan* (draft). Washington, D.C.: U.S. Department of Commerce, PNGV Secretariat.

The White House. 1993. *Historic Partnership Forged with Automakers Aims for Threefold Increase in Fuel Efficiency in as Soon as Ten Years.* Washington, D.C.: The White House.

2

Crosscutting Issues

This chapter addresses crosscutting issues identified by the committee that, in its opinion, require attention. Some were addressed in the Phase 1 and 2 reports (NRC, 2005, 2008). The areas addressed here are (1) program decision making, (2) safety, (3) the balance between "short-term" and "long-term" R&D activities, (4) the interface of plug-in electric vehicles with the nation's electricity delivery system, (5) persisting trends in automotive innovation, and (6) environmental issues associated with different vehicle/energy source pathways. Specific technical areas being addressed by the Partnership are considered in Chapters 3 and 4.

PROGRAM DECISION MAKING

The topics of strategic planning, program management, and decision making within the FreedomCAR and Fuel Partnership are all closely related, and they all critically depend on systems analysis. As described in Chapter 1, the Partnership is a research and development (R&D) program that focuses on critical transportation technology and fuels challenges for vehicles; if successfully met, these challenges could significantly lower U.S. petroleum consumption and greenhouse gas (GHG) emissions. The Partnership's individual technical teams, which include members from the DOE, national laboratories, the automotive OEMs and suppliers, energy companies and power companies, work primarily at the vehicle component level and on the production, distribution, and delivery of hydrogen; in addition, there is recent attention on the interface between the nation's electricity delivery system and the charging of electric vehicles (e.g., plug-in hybrid electric vehicles [PHEVs] or all-electric or battery electric vehicles [BEVs]). There are annual DOE program reviews in addition to many DOE-sponsored conferences and workshops as well as considerable participation in professional society conferences to help keep everyone in the Partnership technical teams well informed. To these teams are added a vehicle systems analysis technical team (VSATT) and a fuel pathway integration technical team (FPITT). This organizational structure is based on project activities that focus on individual technical issues, as well as on total vehicle system integration and the total fuel chain (see Figure 1-1 in Chapter 1). In addition, there is a broader strategic perspective, which the Executive Steering Group (ESG) provides. The system integration and performance issues require a systems analysis approach on several levels, necessitating a variety of systems analysis tools.

In its previous reports, the National Research Council (NRC) recommended substantial activity to develop systems analysis tools to help the Partnership meet its goals. For example, in its first report it was recommended that "an ongoing, integrated, well-to-wheels assessment be made of the Partnership's progress toward its overall objectives" (NRC, 2005, p. 9). In its second report, the committee recommended that "the DOE should accelerate the development and validation of modeling tools that can be used to assess the roles of various propulsion system and vehicle technologies and fuels, and utilize them to determine the impact of the various opportunities on the overall Partnership goals of reducing petroleum use and air pollutant and greenhouse gas emissions" (NRC, 2008, p. 13).

The Partnership has made substantial progress on the development and application of these systems analysis tools. Well-to-wheels analysis (the committee now generally uses the term "source-to-wheels")[1] is now routinely used across the Partnership, and modeling and simulation tools are widely used within the technical teams to support detailed design and analysis as well as target setting. The impact on goals is being assessed by integrating information from various models such as the GHG information from the GREET (Greenhouse Gases, Regulated Emissions, and Energy Use in Transportation) model and market interaction information from national energy sector models (MARKAL [Market Analysis] and NEMS [National Energy Modeling System]). The PSAT (Powertrain Systems Analysis Toolkit) model provides vehicle performance information based on goals and targets. Key program target information for advanced diesel vehicles, hybrid electric vehicles (HEVs), PHEVs, and fuel cell vehicles are included in the PSAT vehicle performance modeling. The hydrogen production and delivery targets are assessed with the H2A (Hydrogen Technology) Production and Delivery models. The MARKAL and NEMS models are used to perform sensitivity analyses of the impacts of meeting or not meeting various targets on the market

[1] The committee chose to use the term "source-to-wheels" instead of "well-to-wheels." In conducting full-fuel-cycle analysis for petroleum-based fuels, the literature has used "well-to-wheels," since petroleum comes from oil wells. However, since transportation energy may now derive from a diverse set of sources, such as solar energy conversion to biomass or coal conversion to electricity, it is more accurate to consider the source of the energy.

shares of various light-duty vehicle technologies, as well as on oil savings and environmental impacts.

Overall, as noted, the development and deployment of systems analysis tools and models at the vehicle and fuel pathway level are impressive, and fully responsive to the committee's specific prior recommendations. However, the systems analysis teams, particularly the VSATT, operate in a support role to the individual technical teams. The application of systems analysis to the overall guidance and management of the Partnership and the determination of technical directions in pursuit of the Partnership's overarching goals relating to national energy policy are much less transparent. In the Phase 2 report (NRC, 2008, p. 30), the committee said that "there is no lack of technical review of the individual program elements, but what is missing is analysis of the quantitative impact on the overall goals of reducing petroleum use and pollutant and greenhouse gas emissions. Tools for estimating this are being worked on: one example is the Macro System Model (MSM) which is scheduled for completion in 2008." As of August 2009, the MSM was reported still to be "under development."

The committee was encouraged to learn at its meeting in October 2009 that the Department of Energy (DOE) has begun using system-level analysis to guide overall program goals and direction, and that sensitivity analysis is being performed on the impact of not meeting different program targets. However, this remains an area in which the committee strongly encourages additional emphasis.

Furthermore, the ESG, charged with overall Partnership guidance, has not met for almost 2 years, leaving an apparent guidance vacuum at the senior leadership level. Although the Partnership has made good progress over this period, it is important that the ESG be fully engaged in the current, ongoing review of the future structure of the Partnership. The committee is assured that this concern is recognized by DOE executive management.

The committee also suggests that the FreedomCAR and Fuel Partnership consider the use of the Oak Ridge National Laboratory (ORNL) consumer choice model to measure the progress of several key advanced vehicles. The reason is that technical progress for advanced vehicles is currently being presented primarily at the subsystem and component levels in a wide variety of units—for example, in terms of fuel economy, range, refueling time, and so on. The lack of a common unit of measure means that the benefits at the subsystem and component levels cannot be combined and compared against cost to get a single value proposition for the collective impact of the advanced technologies on the full vehicle system. The consumer choice model, however, converts the technical advances into the same unit, dollars, thereby allowing the improvements in the value-versus-cost proposition to be estimated.[2]

---

[2] See, for example, D. Greene and L. Zhenhong, "The MA^3T Model: Market Adoption of Advanced Automotive Technologies," Presentation to the committee, December 10, 2009, Washington, D.C.

In summary, the two systems analysis teams have done excellent work and have made great progress at the micro level, but although there are signs of improvement, it is still unclear to the committee how or if this work is being adequately applied at the senior leadership level to guide overall Partnership direction.

SAFETY

Overview

The transition to alternative-fueled vehicles—whether using electricity, biofuels, or hydrogen—will involve new safety challenges that need to be identified and resolved for each alternative. This section on safety emphasizes the safety of hydrogen systems, but the other alternatives also deserve attention.

An exemplary hydrogen safety record will not ensure the success of fuel cell vehicles and other hydrogen technologies under development by the Partnership and the eventual transition to a hydrogen economy; however, a poor safety record may delay or inhibit the widespread use of hydrogen. The goals and objectives of the broad safety portion of the Partnership are to develop practices and procedures that will ensure safety in the operation, handling, and use of hydrogen and hydrogen systems for all DOE-funded projects and to implement these practices and lessons learned to promote the safe use of hydrogen.

The goals and objectives of the narrower codes and standards portion of the program are as follows:

- To perform the underlying research to enable codes and standards to be developed for the safe use of hydrogen in all applications, and
- To facilitate the development and harmonization of domestic and international codes and standards.

The DOE safety, codes, and standards program is focused on hydrogen. Its budget from fiscal year (FY) 2006 to FY 2010 is shown in Table 2-1.

The budget has been robust since FY 2007, but there is a significant reduction for FY 2010. The codes and standards portion is included in the Partnership's codes and standards technical team. The safety part is administered by DOE Headquarters.

The breakdown of the DOE safety, codes, and standards program into six subprogram elements with funding for FY 2009 and FY 2010 is shown in Figure 2-1. The hydrogen codes and standards subprogram focuses on the research and development needed to strengthen the scientific basis for technical requirements incorporated in national and international standards, codes, and regulations. The subprogram also sponsors a national effort by industry, standards, and model-code development organizations and government to prepare, review, and promulgate

TABLE 2-1 U.S. Department of Energy Safety, Codes, and Standards Funding from FY 2006 through FY 2010

Fiscal Year	Funding ($)
2006	4,595,000
2007[a]	13,492,000
2008[a]	15,442,000
2009[b]	12,500,000
2010	8,839,000
Total	54,605,575

[a] FY 2007 and FY 2008 numbers exclude Small Business Innovation Research and Small Business Technology Transfer (SBIR/STTR) funding.

[b] Under the Vehicle Technologies Program budget in FY 2009.

SOURCE: Response from DOE to committee questions, November 23, 2009.

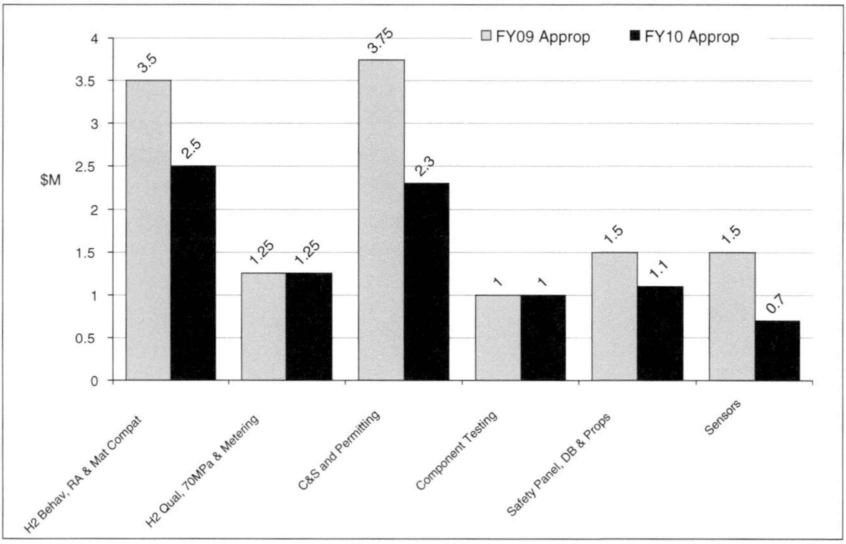

FIGURE 2-1 U.S. Department of Energy safety, codes, and standards budget allocation, FY 2009 and FY 2010 (millions of dollars), broken down by six subprogram elements. The FY 2009 budget was $12.5 million and the FY 2010 budget was $8.8 million.

NOTE: H2 Behav = hydrogen behavior; RA = risk assessment; Mat Compat = materials compatibility; H2 Qual = hydrogen quality; C&S = codes and standards; DB = database; Props = properties. SOURCE: Antonio Ruiz, U.S. Department of Energy, November 2009.

hydrogen codes and standards needed to expedite hydrogen infrastructure development and to help enable the emergence of hydrogen as a significant energy carrier. The overall goal of the safety subprogram is to understand, develop, and promote the practices that will ensure the safe handling, storage, and use of hydrogen. By promoting hydrogen safety procedures, supporting a research program, and developing information resources, the safety subprogram seeks to help form the basis for the safe use of hydrogen as an energy carrier, now and in the future.

The codes and standards portion of the Partnership, which includes the R&D Roadmap and National Template, funds several organizations developing vehicle-level and component-level safety standards. Considerable progress is being made.

The safety part of the program includes the Hydrogen Safety Panel, a Web-based incident reporting system,[3] a bibliographic database,[4] and a Best Practices Web site.[5] There is also an extensive program on unintentional releases of hydrogen and on hydrogen behavior, safety sensors, and the compatibility of materials with hydrogen.

Hydrogen safety activities in the U.S. Department of Transportation (DOT) are partially funded by the DOE. DOT has a hydrogen-related budget of about $14 million, of which about $10 million is spent on the hydrogen bus program in the Federal Transit Administration (FTA). There may be safety lessons learned from the bus program. The majority of the safety-related work is in DOT's National Highway Traffic Safety Administration (NHTSA) and DOT's Pipeline and Hazardous Materials Safety Administration (PHMSA).

Response to Phase 2 Report

The full text of the recommendations discussed in this subsection comes from the Phase 2 report (NRC, 2008) and is reprinted in Appendix D of this report.

- *Recommendation 2-7: DOE should establish a program to address all end-to-end safety aspects.* The response relative to vehicle safety is incomplete and refers to nonexistent DOT standards. Analysis can be done on generic vehicle safety issues as well as the end-to-end fuel supply chain. The DOE could choose a national laboratory to lead this effort.
- *Recommendation 2-8: DOT should develop a long-range hydrogen safety plan.* Apparently this has been done, and these milestones have been integrated into the codes and standards technical team roadmap.

[3] See <www.h2incidents.org>.

[4] See <http://www.hydrogen.energy.gov/biblio_database.html>.

[5] See <www.h2bestpractices.org>.

- *Recommendation 2-9: The codes and standards technical team should update its roadmap to 2015.* This has been done.
- *Recommendation 2-10: DOE should establish a program to collect and analyze safety data on compressed natural gas and hydrogen components, subsystems, vehicles, and fueling stations.* A DOT/DOE workshop to begin this process was held in December 2009. The committee suggests that this work be expanded and continued. This work can provide input to an end-to-end quantitative risk analysis.
- *Recommendation 2-11: DOE should convene a panel of outside experts in the hydrogen compatibility of materials.* The response was very thorough and included interactions with many stakeholders. A review panel of independent experts was planned for FY 2009. This area should continue to be of high priority.
- *Recommendation 2-12: DOE should accelerate work on delayed ignition of unintended hydrogen releases.* The response was very complete and excellent.

In addition to the six recommendations in the Safety section of the Phase 2 report, there was a safety-related recommendation in the Onboard Hydrogen Storage section:

- *Recommendation 3-9: The Partnership should perform studies to determine the risks and consequences of relying on pressurized hydrogen storage. Safety issues should be included in these studies.* The response covers everything except the safety risks and consequences of high-pressure storage. It is suggested that a comprehensive program for compressed hydrogen tank safety be developed and executed. Very little tank-level safety testing has been done, and there are new initiatives to reduce the tank weight and cost, which could influence the tanks' safety performance. An example is the initiative to change the burst margin (discussed below in the subsection entitled "High-Pressure Hydrogen Storage Safety Issues").

Discussion

This discussion on safety addresses four areas. The committee's recommendations are presented at the end of this "Safety" section.

1. End-to-End Safety Analysis for All Vehicle and Fuel Pathways

The Phase 1 report (NRC, 2005) included a safety recommendation for forming a "crosscutting safety technical team." That was not accomplished and thus was amplified in the Phase 2 report, as follows: "DOE should establish a program to address all end-to-end safety aspects" (NRC, 2008, p. 12). That recommendation was partially accepted in the Partnership's responses to the recommendations in the Phase 2 report, although the committee has yet to see any results from that effort.

The Partnership was originally focused primarily on hydrogen/fuel cell vehicles. Now that the program is putting significant emphasis on other propulsion systems and fuels, the safety program should be expanded to cover all of them. The fuel pathways should be examined from source to wheels, and the vehicles studied should include HEVs, PHEVs, biofueled vehicles, and BEVs as well as those powered by hydrogen/fuel cells. The analysis should use a "Life Cycle Assessment" methodology, which is even broader than "cradle-to-grave," as it includes the recycling or reuse of all elements at the end of life (see the section "Environmental Impacts of Alternative Pathways," below, as well as the committee's interim letter report, included as Appendix B in this report).

The six alternate fuels have been defined by the DOE as hydrogen, electricity, natural gas, ethanol, propane (or liquefied petroleum gas [LPG]), and biodiesel. (The committee suggests that the Partnership also add other potential biofuels in addition to ethanol.) The National Renewable Energy Laboratory (NREL) has recently completed a codes and standards "gaps analysis" for each of these fuels, and NREL finds that work needs to be done on many of them, with electricity and hydrogen needing the most attention (Blake et al., 2010).

The demonstrable safety of battery and high-voltage electrical systems is plainly essential for the commercial success of the HEVs, PHEVs, and BEVs.[6]

At the committee's August 2009 meeting, the hydrogen production and delivery technical teams said that they still needed more data on hydrogen behavior. The Partnership should ensure that their needs are documented and addressed. The NREL Wind2H2 demonstration project has also asked for help in identifying and streamlining the codes and standards that would be necessary to deploy all of the components and systems needed for a wind-to-hydrogen system or a wind-to-hydrogen storage-to-electricity power plant. These components and control protocols would include the wind turbines, power electronics, electrolyzers, hydrogen stationary storage, and stationary fuel cells.

2. High-Pressure Hydrogen Storage Safety Issues

For the foreseeable future, compressed hydrogen is the most likely onboard hydrogen storage method for fuel cell vehicles. The need for acceptable vehicle range and trunk space dictates that the pressure vessel accommodate 70 MPa (about 10,000 psi) of gas safely. Containment of the high pressures will likely require pressure vessels to be made of carbon-fiber composite having an interior

[6] For further discussion of battery safety, see in Chapter 3 the section "Electrochemical Energy Storage."

liner made either from a metal (e.g., aluminum) or a polymer (e.g., high-density polyethylene). There are strong motivations for reducing the cost and weight of pressure vessels, and it has recently been proposed to reduce the burst pressure ratio of Type 3 and 4 carbon-fiber tanks from 2.25 (beginning of life) to 1.8 (end of simulated life). Also, compressed-gas tanks are susceptible to fire damage and need to be protected from fire by pressure-relief devices. A new generation of such devices can protect the entire length of the tank from localized fire.

Hydrogen can enter a variety of metals and alloys as H^+ ions and can seriously degrade the structural properties of the metal. High-strength steels and steel welds are particularly susceptible, aluminum much less so. Hydrogen can also diffuse into polymers as H_2 molecules and collect in voids forming blisters and cracks.

Natural gas vehicles use high-pressure gas storage, and many of the hydrogen components and systems are similar. Gathering and analyzing this experience can help ensure that hydrogen vehicles are safer.

3. Emergency Response Issues and Procedures

There are both crash and fire safety issues related to liquid-fueled, electric, and hydrogen vehicles. Some preliminary emergency response guidelines have been developed by the automobile manufacturers (original equipment manufacturers, or OEMs), the California Fuel Cell Partnership, and the DOE's HAMMER (Hazardous Materials Management and Emergency Response) facility. In general, these were developed using commonsense judgments and, of course, have differences. It would be useful to do research and risk analyses to contribute to a better understanding of the most effective strategies. One issue is how to identify the type of vehicle and energy storage method (compressed, liquid, or hydride for hydrogen; or liquid fuel for internal combustion engines [ICEs]; or high-voltage batteries for many vehicle types). The fire-fighting techniques for each of these can be very different. What kind of suppressant should be used (water, foam, CO_2, special hydride powders, etc.)? A range of vehicles should be considered, including passenger vehicles, medium- and heavy-duty vehicles, and both liquid and gaseous hydrogen delivery trailers. This work should be done in conjunction with the emergency response community and with one or more universities that have fire technology programs. Field experiments should be conducted.

4. Lack of Visibility of Department of Transportation Efforts

The DOT parts of the safety program are not visible to the committee. Both the NHTSA and PHMSA have significant roles. The Phase 1 report included a recommendation for getting NHTSA more involved (NRC, 2005). In the Phase 2 report, the committee recommended that "DOT should develop a long-range, comprehensive hydrogen safety plan" and that the DOT milestones should be integrated into the codes and standards technical team roadmap to 2015 (NRC,

2008, p. 12). The committee encourages more visibility of the DOT in the R&D that is being conducted and in the various national and international rule-making efforts.

Appropriate Federal Role

Addressing and ensuring safety is an essential federal role. Most of the safety program would not happen without government funding, and all of the work is appropriate.

Recommendations

Recommendation 2-1. The Partnership should establish a program to address all end-to-end safety aspects in addition to the existing codes and standards work. This work should be based on the pathways work and should include production, distribution, dispensing, and the vehicles. It should apply to all six alternative fuels and their associated vehicle types, including the use of high-voltage electricity on many of these vehicles.

Recommendation 2-2. The Partnership should generate and act on a failure modes and effects analysis of the full pressure vessel assembly, which includes the attached components and the human interface at the pump. Accelerated laboratory tests need to be run to identify failure/degradation modes of the pressure vessel and the mechanisms leading to failure. A nondestructive test program needs to be developed to assess pressure vessel integrity, which should serve both as a tool for quality control and as a means of checking for damage in service. The work on the analysis of worldwide natural gas and hydrogen incidents should continue. An R&D program should be established to develop a new generation of pressure-relief devices that can protect the storage tank from localized fire.

Recommendation 2-3. The hydrogen compatibility (including embrittlement) program should be continued. The Partnership should have experts in hydrogen embrittlement review the operating conditions and materials in the high-pressure delivery and refueling stations for potential problem areas, including welds and nonmetallic materials.

Recommendation 2-4. The Partnership should establish an emergency response R&D program with the involvement of emergency responders and research organizations to do fundamental work on the response to incidents involving alternative fuels. High-voltage batteries and electrical systems should also be included.

Recommendation 2-5. The Partnership should fully integrate the DOT safety efforts into the safety and the codes and standards aspects of the FreedomCAR and Fuel Partnership. All relevant parts of the DOT should be included: those involving passenger vehicles, trucks, the hydrogen bus program, pipelines and hazardous materials, fuel delivery trailers, and others. Alternative fuels should be included. The DOE and the Partnership's Executive Steering Group should consider adding a high-level DOT representative to the ESG.

BALANCE BETWEEN SHORT-TERM AND LONG-TERM ACTIVITIES

In the Phase 1 and Phase 2 reviews (NRC, 2005, 2008), the committee noted that the distribution of funding as well as overall Partnership efforts between short-term and long-term activities seemed to appropriately favor the long-term projects. Indeed, most project efforts seemed to be devoted directly or indirectly to research leading to technologies for achieving a hydrogen production and delivery infrastructure, effective vehicle onboard hydrogen storage, and vehicle fuel cells that could be mass-manufactured at acceptable costs. This type of distribution seemed appropriate to the committee, since the primary justification for government involvement was considered to be the long-term, high-risk, high-payoff type of R&D that probably would not be done without government participation.

Between Phases 2 and 3, major changes took place in the economic and political forces that help shape and direct such issues. The combination of the near collapse of automobile manufacturing in the United States, the economic problems at least partially associated with trade imbalance including the enormous dollar value of imported petroleum, and the growing national and international concerns with greenhouse gases dramatically changed the picture. Whether or not it proves to be altogether true, the perception was, and is, that there are alternative routes for addressing these problems and that the pursuit of fuel cell vehicles, utilizing hydrogen fuel, represents the longer-term option.

The committee agrees that there are other options, and it identifies three primary alternative routes to reducing U.S. petroleum consumption: (1) vehicles utilizing hydrogen fuel and fuel cell power plants, (2) vehicles with internal combustion engines using biofuels, and (3) greater electrification of the vehicles (e.g., PHEVs or BEVs), thus shifting part of the transportation energy from petroleum to grid electricity. The committee also agrees that of the three, the hydrogen and fuel cells option is expected to be the longer-term option. Even so, the other options also have major issues to be resolved. For example, to make PHEVs, and especially BEVs, practical and affordable requires better battery technologies and lower costs than are currently available. The same is true for biofuels, for which much-improved processes must be developed and abundant renewable feedstocks identified in order to avoid fuels competing with foods. These issues and many others, and given the present circumstances, indicate a need for government-assisted R&D.

With this changed background, the committee believes that it is proper to shift a larger share of the Patnership's efforts and funding to R&D for nearer-term technologies. However, the committee also believes that R&D for the longer-term technologies, especially hydrogen and fuel cells, should not be abandoned but should also be continued. Especially with the technical uncertainties associated with any of the technologies being pursued and the unknowns facing the vehicle markets, it is extremely important to have a reasonably balanced portfolio of both short-term and long-term options. The present projected distribution of funds (see Chapter 5) accomplishes both increased efforts for nearer-term technologies and the continuation of an acceptable level of efforts for the longer-term technologies, and therefore the committee believes that this distribution is generally reasonably balanced and appropriate. However, there are technology areas (see Chapters 3 and 4) in which the committee recommends that some increased efforts should be considered.

BATTERY ELECTRIC AND PLUG-IN HYBRID ELECTRIC VEHICLES AND THE U.S. ELECTRIC GRID

In view of the recent formation of the grid interaction technical team and the recent policy and commercial emphasis on PHEVs and BEVs, the committee reviewed the following: (1) the ability of the electric grid to support the entry of the PHEV and BEV and the implications for greenhouse gas reduction, (2) the interface between "smart grid" technologies that manage energy use at the consumers' premises and on-vehicle recharging and energy management systems, and (3) the reuse of spent batteries from PHEVs and BEVs for U.S. electric grid load management, including the balancing of loads from grid-interactive, renewable energy sources.

Electric Grid: Adequacy and Consequences

Numerous automotive OEMs have scheduled plug-in vehicles, either PHEVs or BEVs, for market entry over the next several years.[7] Nevertheless, the impact of these vehicles on the electric grid is not likely to be immediate in the absence of strong market-forcing policies by the federal government. The National Academies' *America's Energy Future* study estimates that the following deployment rates are plausible (NAS/NAE/NRC, 2009):

- PHEVs could account for 1 to 3 percent of the new-vehicle market by 2020, and 7 to 15 percent by 2035;
- BEVs could account for 0 to 2 percent of the new-vehicle market by 2020 and 3 to 10 percent by 2035.[8]

[7] A partial list of OEM-announced plans for the North American market includes BMW, BYD, Fisker, Ford, GM, Honda, Hyundai, Mitsubishi, Nissan, Tesla, and Toyota (NRC, 2009).

[8] These rates are estimated for each vehicle type independently and do not imply that both will occur simultaneously.

To be sure, more aggressive penetration rates can be described, but the committee has not assumed that these will occur. The bottom line is that in either case, a shift to being a significant percentage of new-vehicle sales involves major transitions that would take decades.

Based on the penetration rates indicated above, the aggregate U.S. electric infrastructure seems quite capable of accommodating the market penetration of BEVs or PHEVs. This is because the aggregate demand that such vehicles could place on the electric infrastructure is small relative to the generating capacity of that infrastructure. For example, 1 million PHEVs charging an average of 3 kWh[9] every day for a year would require only about 1 million megawatt-hours (MWh). In contrast, the national electric infrastructure generated 4,157 million net megawatt-hours in 2007 (EIA, 2009). Thus, an analysis by the Pacific Northwest National Laboratory estimated that a PHEV fleet equal in size to 84 percent of all cars and light trucks on the road in 2001 could be charged during off-peak times without building new electricity generation capacity (PNNL, 2007).

However, the aggregate data provide an incomplete guide to policy. First, local grid circuits might become overloaded if the responsible utility fails to anticipate the new demands and/or if the local rate commission fails to provide adequate cost recovery. Furthermore, electric utilities have successfully dealt with the increased loads needed for air-conditioning systems even though these load the grid during times of peak demand. But such increased load during times of peak demand could occur for any new electric load and is not a unique characteristic of BEV/PHEV deployment. More important for policy purposes is the tension between (1) ubiquitous charging opportunities, which would accelerate BEV/PHEV market penetration by relieving consumers of the "range anxiety" widely noted to inhibit electric vehicle (BEV) purchases; and (2) the environmental and cost consequences of recharging vehicles at any time convenient to the driver.

A recent NRC study examined these consequences, and the committee has drawn extensively on that analysis (NRC, 2009): *Transitions to Alternative Transportation Technologies—Plug-in Hybrid Electric Vehicles* examined market penetration rates for PHEVs far exceeding those used in the *America's Energy Future* study cited above (NAS/NAE/NRC, 2009). Even under these circumstances, the NRC analysis agreed with previous studies that the grid capacity is likely to remain adequate for the foreseeable future as long as vehicle charging is during off-peak times. But charging during peak hours raises issues of cost and grid reliability, while charging at any time raises questions of greenhouse gas emissions. The Institute of Electrical and Electronics Engineers (IEEE) P1809 standards committee is working on standards for electric grid-to-vehicle charging.

---

[9] A PHEV-10, which has an all-electric range of 10 miles, has storage capacity of about 2 kWh; a PHEV-40, which has an all-electric range of 40 miles, has a storage capacity of about 8 kWh. Thus, an estimate of 3 kWh for a mixed fleet seems reasonable (NRC, 2009).

Cost and Grid Reliability

Recent analyses by the North American Electric Reliability Corporation (NERC) raised concerns about the reliability of the electric power system, especially during peak hours when projected increases in demand over the 10-year NERC planning horizon exceed currently planned capacity additions (NERC, 2008). Charging PHEVs or BEVs during peak hours, though desirable from a consumer perspective, could add to the prospective shortfall in peak capacity. For example, a study by Southern California Edison concluded that PHEVs could account for as much as 11 percent of its system load by 2020, which could increase peak loads by several thousand megawatts if PHEV charging is not properly managed (NRC, 2009).

Greenhouse Gas Implications

More than 70 percent of the net electricity generation by the U.S. power sector derives from fossil fuels, and hence causes GHG emissions. Assuming that these emissions continue indefinitely, the NRC (2009) analysis showed that the PHEV could still offer modest advantages over a highly efficient HEV beginning around 2035. The PHEV shows a marked advantage over a reference case, conventional ICE vehicles with modest efficiency improvements (NRC, 2009). For a BEV, the corresponding advantage in CO_2 reduction would be greater.[10]

Of course, reducing the carbon footprint of the electric grid would increase this advantage even further. For example, a joint analysis by the Electric Power Research Institute (EPRI) and the Natural Resources Defense Council (NRDC) explored the relationship between the grid and the PHEV using nine modeling scenarios for reduced CO_2 emissions from U.S. electric power generation (EPRI/NRDC, 2007). EPRI and NRDC concluded that all nine scenarios showed significant GHG reductions attributable to PHEV fleet penetration. According to their models, cumulative GHG savings from 2010 to 2050 could range from 3.4 to 10.3 billion metric tons (MT) of CO_2. In contrast, current CO_2 emissions from gasoline used in transportation are currently about 1.2 billion MT per year (NRC, 2009). An NRC committee, however, cautioned that the climate benefits of PHEVs are "small unless the grid is decarbonized with renewable energy, nuclear plants or fossil fuel fired plants equipped with carbon capture and storage technologies" (NRC, 2009, p. 5).

[10] The NRC assessment of the PHEV (NRC, 2009, p. 18) offered some interesting comparisons. "CO_2 emissions by U.S. electric generators and combined heat and power facilities in 2007 were 2,517 million metric tons (EIA, 2009), or an average of about 1.3 pounds of CO_2 per kWh." One kilowatt-hour will take a small BEV about 5 miles. Over the same distance, a typical gasoline-powered car achieving 30 miles per gallon (mpg) would emit about 3 pounds of CO_2. An HEV at 50 mpg would release about 2 pounds.

Vehicle-Grid Interface

A variety of new companies is entering the local grid market to supply energy management technologies that could influence the recharging and use of BEVs and PHEVs.

Smart Grid

In February 2009, Google announced its entry into the smart-microgrid market with a Web application that displays in real time the home energy consumption of each appliance—and vehicle battery being charged. The software uses "smart" meters that can communicate home energy consumption back to utilities every few minutes. In recent months, Microsoft, Verizon, and AT&T have made similar announcements.

Widely accepted, open-architecture standards for information exchange between vehicles and the electric grid, especially to local smart grids, are essential for the deployment of all plug-in vehicles, whether PHEVs or BEVs. The open-architecture nature of these standards could allow a wide range of information and energy management system developers to enter the market quickly and efficiently. In addition, open-architecture standards could connect vehicle charging with market opportunities for renewable energy, perhaps as a distributed resource. Much progress has been made in developing these standards to serve a variety of vehicle-grid communication purposes. IEEE Draft Standard P1901 for "Broadband over Power Line Networks: Medium Access Control and Physical Layer Specifications" was released for sale to the public at the beginning of 2010 (IEEE, 2010).

Battery Charging and Discharging

Charging parameters for PHEVs and BEVs are dependent on the charge power, charging time, and the size and type of the battery. Charging time varies depending on the distance traveled and on the charger type available. The most common charger available in homes is the Level 1, 120 V AC (volts alternating current), 15 A (amps) (12 A usable) or 20 A (16 A usable). Thus a completely depleted 40-mile PHEV would be charged by this charger in 5 to 8 hours. A Level 2 power outlet of 240 V AC and 40 A circuits would charge the same vehicle in about 1 hour. Level 3, or "fast charging" stations that use 480 V AC and up to 100 kW power output are being considered for commercial or public application; these would charge this vehicle in significantly less time. Lithium-ion (Li-ion) batteries being considered for PHEV and BEV applications can easily be charged with Level 1 and 2 chargers but would require special attention, particularly thermal management, during Level 3 (fast charging). However, the general availability of fast chargers would remove the anxiety of running out of energy during longer drives for BEV customers.

The reuse of batteries from other sources to replace failed battery modules in high-voltage strings (of batteries) has been evaluated many times in past. For example, an Electricity Advisory Committee (EAC) report issued in December 2008 discusses the use of energy stored in hybrid electric vehicle batteries to manage the grid (EAC, 2008). In Section 4.2, Phase 3, EAC (2008, p. 21) states: "Control of the bidirectional electric flow could include payments to owners for use of their automobile batteries for load leveling or regulation and for spinning reserve (the cashback hybrid incentive)." Although BEV/PHEV battery charging may have some benefits in electrical grid management, the highest priority must be given to maintaining the safety and reliability of the battery, the vehicle, and the occupants of the home or other facility where the battery is being charged. These considerations require that battery-charging decisions be reviewed and accepted on the basis of battery data continually taken and stored onboard the vehicle by the intelligent battery charger. This is, of course, no different from charging a portable computer that has an intelligent battery charger, except that each vehicle battery will be larger, more powerful, and more expensive than a small personal computer battery. It is thus important to understand that the use of vehicle battery charging to optimize utility grid loads must be limited by these considerations. More-detailed models of grid use that include specific battery-charging requirements based on external factors such as climate and vehicle usage may be required for an understanding of the full impact on the electric utility grid as the electric vehicle market increases.

The discharging of BEV/PHEV batteries for grid management while the batteries are still on the vehicles also raises issues. Battery performance and life are functions of the number and depth of discharges. Using an expensive battery developed for vehicle propulsion for peak shaving of the utility grid would cause much shorter battery life as well as additional maintenance problems in the vehicle. The effect on battery life and its replacement costs should be analyzed and compared with other methods of utility peak shaving, including large battery systems specifically designed for this purpose.

Other Grid Interface Issues

The expense of installing recharging circuits at the premises of customers who do not have them (townhouse or apartment dwellers, for example) or in public places (theaters or shopping malls, for example) could raise a cost barrier in some cases (NRC, 2009). However, commercial markets for these services are well established, and the value that precompetitive research might add is unclear.

Reuse of Spent Vehicle Batteries

The storage batteries used in PHEVs and BEVs are considered worn out when they are no longer able to deliver at least 80 percent of their rated capacity under the conditions of use in a specific application. Examples include use in telecommunications standby power installations, uninterruptible power supply (UPS), and utility safety systems.

Battery manufacturers discourage and generally fail to provide warranties in these cases, because resulting imbalances in battery wear across a string often shorten the life of other batteries and hasten string failure. The reuse of PHEV and BEV batteries will make sense only if the new application can make use of battery capacity at rates where significantly more than 80 percent of the battery capacity is still available.

To ensure consumer safety, batteries should be reused only under strict regulations that prohibit their reuse by vehicle customers in unregulated environments. The reuse of spent storage batteries from PHEVs and BEVs will require the development of facilities with trained personnel who test and match the performance of the batteries using appropriate equipment and standard tests and protocols. Creating such a capability will require significant cooperation between battery manufacturers and users for the development of appropriate standards that are based on the state of health of PHEV and BEV batteries. New knowledge must be also acquired regarding battery failure modes as a function of environmental factors, vehicle design, and use modes.

The reuse of PHEV and BEV batteries will thus require significant initial labor and infrastructure expenditures in cooperative efforts with battery suppliers. If this effort is successful, ongoing costs of battery testing and redeployment may still exceed the value of the reused batteries. Past experience with battery reuse provides no guarantee of success in developing a viable reuse program or warranty support from battery vendors. The net effect of battery reuse would be small and depends on many factors that are not known at this time. The first phase of this program should therefore be a detailed analysis of the costs and benefits of this activity versus the costs and benefits of developing a battery-recycling protocol and system. There are recycling plans for consumer Li-ion batteries, but the quantity of batteries and materials to be recycled would be significantly larger for automotive batteries (even at 2 percent penetration). If the first phase of the study shows that recycling may be a viable option for battery reuse, it should be evaluated further.

Recommendations

Recommendation 2-6. The grid interaction technical team should work with state utility regulatory authorities, perhaps through the National Association of Regulatory Utility Commissioners, to ensure that the incentives provided by state regulations mesh well with the national interest in vehicle deployment, reduced oil consumption, and lower greenhouse gas emissions.

Recommendation 2-7. The grid interaction technical team should continue to encourage and, where appropriate, facilitate the ongoing development of open-architecture standards for smart-vehicle/smart-grid interconnections currently being developed by the Institute of Electrical and Electronics Engineers and the Society of Automotive Engineers. In doing so, the technical team should encourage participation from the purveyors of smart-grid systems and battery suppliers as well as from the electric utility industry.

Recommendation 2-8. Standards for the reuse of electric vehicle batteries should be developed under leadership of the grid interaction technical team, and training materials for the use of these standards should be developed in parallel.

PERSISTING TRENDS IN AUTOMOTIVE INNOVATION: IMPLICATIONS FOR THE FREEDOMCAR AND FUEL PARTNERSHIP

From a strategic perspective, the public issues that motivate the FreedomCAR and Fuel Partnership, chiefly energy and the environment, should be viewed within the context of several persisting trends that will influence the pace and direction of innovation in automobiles and fuels. Some of the most relevant of these are also the most familiar, and thus their implications for the Partnership can easily be overlooked. Here, two of the most relevant persisting trends are briefly analyzed.

Consumer Preferences for Vehicle Cost, Performance, and Safety

Surveys suggest that the total cost of ownership will remain the single most important decision criterion in auto purchases (see, e.g., Oliver Wyman, 2005). In contrast, fuel economy will rise and fall as a decision criterion for consumer purchases as oil prices rise and fall. This implies that vehicles that offer superior fuel economy must not do so at the expense of other attributes desired by the consumer, such as safety and affordability.

Research and innovation are essential, both for fuel economy and for lowering the costs and improving the safety performance of the fuel-saving vehicles. Notwithstanding, economic pressures remain likely to constrain the level of private R&D activities, affecting (1) the amount that can be spent, (2) the strategic purpose of the research efforts, and (3) whether the available R&D funds are spent internally or outsourced. At the same time, manufacturing innovation can improve the ability of automotive OEMs and suppliers to respond better to volatile consumer preferences and to lower vehicle cost. Hence, manufacturing innovation and an efficient innovation process for gaining access to advanced technology, both for vehicles and for the manufacturing processes by which they are made, are becoming essential for competitive success. Vehicle technology is addressed first.

Sources of Vehicle Technology

Increasingly, the sources of advanced technology will arise from outside the traditional automotive OEMs and their top-tier suppliers. Currently, as much as 70 percent of the value added for a new vehicle derives from the supplier networks of the OEM (Dyer and Nobeoka, 2000). For example, microelectronic devices can contribute strongly to the vehicle attributes preferred by purchasers (especially energy-saving and emissions improvements) as well as to cost reductions in the vehicles themselves. Yet many of the most compelling of these devices and software originate from outside the automotive OEMs and suppliers, which would benefit from cost-effective access to these. Hence the management of innovation networks will be central for OEMs to access new technologies and innovative ideas.

In contrast, the auto companies will continue to be a leading source of innovation in whole-vehicle systems—for example, the power train—which will continue to evolve for improved performance and energy savings. But even in power trains, value might be gained from access to technologies originating from outside the traditional industry.

These persistent trends suggest that the FreedomCAR and Fuel Partnership should think beyond R&D and revisit its innovation strategies. Among the strategic issues to consider, the committee suggests the following in the form of recommendations.

Recommendations

Manufacturing

Recommedation 2-9. The Partnership should consider including manufacturing processes among the precompetitive R&D programs. Because its funding originates in the United States, the Partnership should emphasize the technologies and methods most capable of realizing advanced vehicle production in the United States, to the extent that this is feasible.

Standards

Recommendation 2-10. As the basic platform of the automobile becomes more modular, interface standards will be required to enable greater competition among technology alternatives. While specific interface standards have been discussed elsewhere in this report, the Partnership should also consider conducting a more general review of areas in which industry-wide standards could accelerate the pace of innovation and lower its cost.

Inclusive Innovation Architecture

Recommendation 2-11. The Partnership should seek out and implement methods to allow new, nontraditional suppliers—especially, emerging entrepreneurial companies—to participate in the innovation process. The Small Business Innovation Research (SBIR) program can become a highly productive source of innovation, and the Partnership should review its linkages with this program and strengthen them where appropriate.

ENVIRONMENTAL IMPACTS OF ALTERNATIVE PATHWAYS

As noted in both the Phase 1 and Phase 2 reports, it is important to understand and analyze the environmental implications of the full fuel cycle, from source to end use, of a hydrogen economy (NRC, 2005, 2008). Such full-fuel-cycle analyses are also important for any of the other energy source/vehicle combinations (e.g., biofuels for ICE or hybrid vehicles, electricity for PHEVs or BEVs) that are being developed that can potentially reduce the consumption of petroleum and greenhouse gas emissions from light-duty vehicles.

To motivate full-fuel-cycle analysis, note that systems-wide effects associated with various technologies potentially lead to unforeseen and important effects on the environment. One type of unforeseen effect would occur when a technology had superior performance in one phase of the supply chain (e.g., a more efficient engine) but caused changes elsewhere in the supply chain (e.g., in producing fuels for the new engine) that reduced or even canceled the benefits. To cite some examples, there is a debate about whether corn-based ethanol actually reduces fossil fuel use and carbon emissions (Farrell et al., 2006). The high energy cost for liquefying hydrogen inflicts a substantial penalty on the source-to-wheel efficiency of a fuel cell vehicle (Bossel, 2006). A second type of unforeseen effect occurs when a technology designed to mitigate a targeted environmental issue induces other types of environmental impacts. The increased cultivation of corn to produce ethanol, for example, leads to an increased use of fertilizer, which could increase the runoff of nitrogen and phosphorus, in turn widening the areal extent of the dead zone in the Gulf of Mexico (Donner and Kuchari, 2008). Some battery technologies contain toxic materials (such as lead) that, as extensive regulation mandating the collection and recycling of lead-acid batteries shows, could pose an environmental hazard unless properly managed (Lave et al., 1995). Fuel cells and batteries may rely on rare materials such as platinum, possibly inducing resource scarcity (Gordon et al., 2006). In addition, there is increased concern over the life-cycle water use of new energy technologies (Webber, 2007). The committee is not claiming that these undesirable effects will happen but that prior careful and complete analyses are needed to ensure that they do not.

The assessment and management of technology systems constitute a rapidly growing area that is being formalized with research disciplines, journals, and professional societies. "Industrial ecology"[11] is an umbrella concept involving a holistic environmental view of industrial systems, including strategies such as industrial symbiosis to maximize the reuse and recycling of resources (Graedel and Allenby, 2009). Materials flow analysis (MFA) is used in industrial ecology to characterize physical flows in industrial systems (NRC, 2004). Life cycle assessment (LCA)[12] is a set of methods and tools to assess supply chain impacts of technology (Hendrickson et al., 2006). LCA is sometimes termed well-to-wheels analysis (more appropriately termed source-to-wheels, as adopted in this report) in the context of vehicle systems.[13] Industrial ecology, MFA, and LCA are increasingly used in policy. For example, life cycle assessment of fuel systems has been explicitly included in national policy in the biofuels arena in the Energy Independence and Security Act of 2007 (Public Law 110-140, H.R. 6).

The FreedomCAR and Fuel Partnership has been working to address supply chain effects of technology. Notably the GREET model for source-to-wheels analysis of transport systems developed at the Argonne National Laboratory is well known, and its functionality and coverage are being expanded (ANL, 2009). The systems analysis team is undertaking useful analyses of environmental issues such as the water use and resource constraints associated with alternate fuel pathways.

To understand the impacts across the full fuel cycle of producing, distributing, and using hydrogen, the Phase 1 report recommended that the DOE, in collaboration with the Environmental Protection Agency, should systematically identify and examine the possible long-term ecological and environmental effects of the large-scale use and production of hydrogen from various energy sources. These direct and indirect effects should include effects on land, water, and the atmosphere. In its response dated April 2, 2009, to the recommendations in the Phase 2 report, the DOE concurred with this recommendation (DOE, 2009, p. 23); its Office of Science (SC) is developing a fundamental understanding of the processes involved in the biogeochemical cycling of atmospheric hydrogen. This knowledge will make it possible to perform a comprehensive assessment of the environmental impact of the release of hydrogen to the atmosphere from large-scale use and production. In fact, the DOE supported two studies on the environmental impacts of a hydrogen economy, with a primary focus on atmospheric impacts; the results of the studies

---

[11] The International Society for Industrial Ecology (ISIE) is the main professional society for industrial ecology.

[12] There are a number of professional societies relating to LCA, including ISIE. The American Center for Life Cycle Assessment organizes an annual LCA conference in the United States, InLCA.

[13] However, many analysts do not consider the impacts of the recycling of materials in what they would refer to as "well-to-wheels" analysis, and some analysts now use the term "cradle-to-grave" to consider not only the impacts of getting energy through the full fuel cycles but also the impacts of recycling materials as well as the eventual disposal of unused materials. Hence, LCA in its fullest meaning would imply a cradle-to-grave analysis.

were presented to the committee at its meeting on December 10, 2009.[14] Some of the conclusions of these studies are as follows:

- The adoption of a hydrogen-fuel-cell-based transportation sector would dramatically improve tropospheric and regional air quality. Although there are some concerns about a decrease in stratospheric ozone, impacts would be greatly reduced in most cases.
- Adaptive soil microbial uptake of hydrogen could provide a powerful negative feedback to future increased hydrogen concentrations and mitigate any adverse impacts on stratospheric ozone.
- Atmospheric hydrogen effects on structures and embrittlement are not likely to be important.
- If hydrogen is produced with carbon-free processes, there will be substantial reductions in future atmospheric CO_2 concentrations, as well as emissions of importance such as oxides of nitrogen and nonmethane organic gases.

While the work on systems effects of technology under the auspices of the DOE is clearly valuable, the committee has several observations on how efforts addressing environmental impacts of fuel pathways could be improved. First, it is not clear that previous work done by outside groups (e.g., associated with professional societies for industrial ecology and life cycle assessments) is being fully utilized. As mentioned above, there is a significant and growing literature in this field, and it would serve the DOE well to base its efforts in the context of this other work.

Secondly, the committee observed a need for stronger integration between systems analyses and the technical teams. Systems analysis can inform technical teams about targets and choices of what technologies are developed, and, conversely, technical teams can provide information to systems analysis on what technologies need to be evaluated. This feedback loop could be strengthened. For example, the GREET model indicates that the energy intensity of hydrogen liquefaction severely penalizes the source-to-wheels efficiency of fuel cell vehicles powered through this route (Wang, 2002). Energy efficiency, however, is not among the explicit targets for liquefaction technology set for the hydrogen production and the hydrogen delivery technical team (DOE, 2009).[15]

---

[14] D. Wuebbles, "Evaluation of the Potential Environmental Impacts from Large-Scale Use and Production of Hydrogen in Energy and Transportation Applications," and T. Grieb, "Potential Environmental Impacts of Hydrogen-Based Transportation and Power Systems," Presentations to the committee, December 10, 2009, Washington, D.C.

[15] A. Sudik et al., "Hydrogen Storage Joint Technical Team," Presentation to the committee, August 5, 2009, Southfield, Michigan.

Recommendation 2-12. The Partnership should undertake a review of the state of methods and case studies that have been carried out on environmental impacts related to the technologies under development. This review would answer some remaining open questions and help direct systems studies so as to maximize their efforts to characterize the environmental impacts of different fuel pathways.

Recommendation 2-13. The Partnership should strengthen the links between the systems analysis teams and the technical teams. In particular, technological goals and targets should include consideration of priorities established in systems analysis, and systems analysis should be conducted on emerging technologies identified by the technical teams.

Recommendation 2-14. The Partnership should consider incorporating the broader scope of a "cradle-to-grave" analysis rather than a "source (well)-to-wheels" approach in program planning from production to recycling in order to better consider total energy consumption, total emissions, and the total environmental impact of various energy/vehicle pathways and technologies.

REFERENCES

ANL (Argonne National Laboratory). 2009. *Greenhouse Gases, Regulated Emissions, and Energy Use in Transportation (GREET) Model.* Argonne, Illinois. Available on the Web at <http://www.transportation.anl.gov/modeling_simulation/GREET/>.

Blake, C., W. Buttner, and C. Rivkin. 2010. *Vehicle Codes and Standards: Overview and Gap Analysis.* NREL/TP-560-47336. February. Available on the Web at <www.nrel.gov/publications>.

Bossel, U. 2006. Does a hydrogen economy make sense? *Proceedings of the IEEE* 94(10):1826-1837.

DOE (U.S. Department of Energy). 2009. *Actions and Evidence Report, April 2.* Submitted to the National Research Council (NRC) Committee on Review of the FreedomCAR and Fuel Research Program, Phase 3, documenting responses by DOE to recommendations in the Phase 2 report [see NRC, 2008].

Donner, S.D., and C.J. Kuchari. 2008. Corn-based ethanol production compromises goal of reducing nitrogen export by the Mississippi River. *Proceedings of the National Academy of Sciences* 105(11):4513-4518.

Dyer, J.H., and K. Nobeoka. 2000. Creating and managing a high-performance knowledge sharing network: The Toyota case. *Strategic Management Journal*, Vol. 21: pp. 346-367.

EAC (Electricity Advisory Committee). 2008. *Bottling Electricity: Storage as a Strategic Tool for Managing Variability and Capacity in the Modern Grid.* December. Available on the Web at <www.oe.energy.gov/eac.htm>.

EIA (Energy Information Administration). 2009. *Electric Power Industry 2007: Year in Review.* January 21. Washington, D.C.

EPRI/NRDC (Electric Power Research Institute/Natural Resources Defense Council). 2007. *Environmental Assessment of Plug-in Hybrid Electric Vehicles. Volume 1: Nationwide Greenhouse Gas Emissions.* July. Available on the Web at <http://my.epri.com/portal/server.pt?open=514&objID=223132&mode=2>.

Farrell, A.E., R.J. Plevin, B.T. Turner, A.D. Jones, M. O'Hare, and D.M. Kammen. 2006. Ethanol can contribute to energy and environmental goals. *Science* 311:506-508.

Gordon, R.B., M. Bertram, and T.E. Graedel. 2006. Metal stocks and sustainability. *Proceedings of the National Academy of Sciences* 103(5):1209-1214.

Graedel, T.E., and B.R. Allenby. 2009. *Industrial Ecology and Sustainable Engineering.* Prentice Hall.

Hendrickson, C.T., L. Lave, and H.S. Matthews. 2006. *Environmental Life Cycle Assessment of Goods and Services: An Input-Output Approach.* Washington, D.C.: Resources for the Future Press.

IEEE (Institute of Electrical and Electronics Engineers). 2010. Draft Standard P1901 for "Broadband over Power Line Networks: Medium Access Control and Physical Layer Specifications." Available on the Web at <http://ieeestandards.org/ct.html?rtr=on&s=8nv.1asmo.2vi2.89py.8sly.jenu.coes>.

Lave, L.B., C.T. Hendrickson, and F. McMichael. 1995. Environmental implications of electric cars. *Science* 268(5213):993-995.

NAS/NAE/NRC (National Academy of Sciences/National Academy of Engineering/National Research Council). 2009. *America's Energy Future: Technology and Transformation.* Washington, D.C.: The National Academies Press.

NERC (North American Electric Reliability Corporation). 2008. *Long-Term Reliability Assessment: 2008-2017.* October. Available on the Web at <http://www.nerc.com/files/LTRA2008v1_2.pdf>.

NRC (National Research Council). 2004. *Materials Count: The Case for Material Flows Analysis.* Washington, D.C.: The National Academies Press.

NRC. 2005. *Review of the Research Program of the FreedomCAR and Fuel Partnership, First Report.* Washington, D.C.: The National Academies Press.

NRC. 2008. *Review of the Research Program of the FreedomCAR and Fuel Partnership, Second Report.* Washington, D.C.: The National Academies Press.

NRC. 2009. *Transitions to Alternative Transportation Technologies—Plug-in Hybrid Electric Vehicles.* Washington, D.C.: The National Academies Press.

Oliver Wyman (formerly Mercer Management Consulting). 2005. *Car Innovation 2015.* Available on the Web at <http://www.car-innovation.com/study-content.html>.

PNNL (Pacific Northwest National Laboratory). 2007. *Impacts Assessment of Plug-in Hybrid Vehicles on Electric Utilities and Regional U.S. Power Grids. Part 1: Technical Analysis.* Available on the Web at <http://www.ferc.gov/about/com-mem/wellinghoff/5-24-07-technical-analy-wellinghoff.pdf>.

Wang, M. 2002. "Fuel Choices for Fuel-Cell Vehicles: Well-to-Wheels Energy and Emission Impacts." 2002 Fuel Cell Seminar, Palm Springs, California, November 18-21.

Webber, M. 2007. The water intensity of the transitional hydrogen economy. *Environmental Research Letters* 2:1-7.

3

Vehicle Subsystems

The long-range goals of the FreedomCAR and Fuel Partnership—to transition to a transportation system that uses sustainable energy resources and produces minimal criteria or net carbon emissions on a life-cycle or source-to-wheels basis—are extremely ambitious. The difficulties are compounded when the additional constraints associated with the Partnership are imposed: energy freedom, environmental freedom, and vehicle freedom. These goals and associated constraints effectively eliminate the continued simple evolution of the gasoline-fueled internal combustion engine (ICE) vehicle as a possible answer. "Sustainable energy resources" and "energy freedom" both suggest non-petroleum-based alternative fuels or electricity. The emphasis on "net carbon emissions" and "environmental freedom" suggests that carbon dioxide (CO_2) and other emissions from the production and consumption of alternative fuels or electricity should be reduced, through highly efficient processes, to minimize adverse environmental effects. Finally, "vehicle freedom" implies that the fuel and onboard energy conversion systems should not limit the options and choice that buyers expect to have available in their personal vehicles. These goals, if attained, are likely to require new transportation energy carriers (fuel[s] and/or electricity) utilized in more efficient power plants in lighter vehicles having reduced power requirements and equivalent utility and safety.

This chapter discusses the vehicle systems technology areas that the Partnership is addressing in its research and development (R&D) programs, which include the following: (1) advanced combustion, emission control, and fuels for ICEs; (2) fuel cells; (3) hydrogen storage on the vehicle; (4) electrochemical energy storage or technologies for storing electricity onboard a vehicle; (5) electrical propulsion systems; and (6) materials for reducing the weight of the vehicle. The reader is referred to the presentations from the Partnership to the committee on the various technical areas: these can all be found in the project's public access file, available through the National Academies Public Access Records Office. Chapter 4 will address issues associated with hydrogen and biomass-based fuels.

ADVANCED COMBUSTION, EMISSIONS CONTROL, AND HYDROCARBON FUELS

Introduction

Steady progress is being made in the advancement of power plants that rely on energy carriers other than liquid hydrocarbon (HC) fuels. However, one unique characteristic of mobility applications is that the energy being supplied to the power plant needs to be carried around with the vehicle. As weight and volume are important parameters in vehicle design and function, it is critical to have the highest possible energy per unit of mass and per unit of volume within the vehicle's fuel system. Here, the fuel system includes all aspects of carrying the energy on the vehicle—that is, the fuel tank or containment system (battery pack, or hydride material) and supporting structures are included in this weight and volume assessment. On this basis, liquid HC fuels are very effective energy carriers for mobility systems.

Using the metrics of energy density (watt-hour per liter [Wh/L]) and specific energy (watt-hour per kilogram [Wh/kg]) of a vehicle's complete fuel system highlights differences compared to conventional vehicles and the challenges of implementing alternative energy carriers to mobility systems. When one makes these comparisons, it is important to consider not only the energy density of the vehicle's fuel system but also the efficiency of converting the energy carried on the vehicle to motive power (power that causes motion) at the wheels of the vehicle.

Liquid HC fuels have very high energy density and specific energy relative to batteries and hydrogen systems, but the efficiency of the ICE is typically lower than that of systems using electric motors and power electronics and fuel cell systems. Thus the concentrated effort to improve the engine and power-train efficiency is easily understood. However, the energy density and specific energy of liquid HC fuels is so great that even considering these efficiency differences, a typical vehicle carrying a liquid HC will have significantly higher capability than that of an electric or hydrogen-powered vehicle in terms of deliverable work to the wheels per unit of mass and volume of vehicle energy storage onboard the vehicle.

For example, comparing an ICE with an efficiency of 40 percent to a hydrogen fuel cell vehicle (HFCV)[1] with an overall power-train efficiency of 65 percent results in a work capacity of the liquid-fueled ICE vehicle that is approximately

---

[1] It has been assumed that the 2015 hydrogen storage targets of 1,300 Wh/L and 1,800 Wh/kg have been met in performing this analysis.

4.5 times higher per unit of volume of "fuel storage" and approximately 4 times higher per unit of mass of "fuel storage" than those of the HFCV. It seems likely that there will be certain applications, such as extended operation at higher loads or very long range transport, that will favor using a liquid HC as the on-vehicle energy carrier.

In addition, as new power plants with alternative energy carriers are developed, produced, and introduced into the market, there will be a significant time delay associated with their market penetration. As noted in Chapter 1, in the United States the vehicle fleet turnover in recent years is estimated to be about 15 years.[2] Consequently the turnover time for completely new vehicle architectures to achieve significant market penetration will be measured in multiple decades (Bandivadekar et al., 2008; Weiss et al., 2000). During this transition the dominant power plant for mobility systems will continue to be ICE vehicles fueled with a hydrocarbon fuel (e.g., gasoline, diesel fuel, or biofuel).[3]

Consequently, it is important to maintain an active ICE and liquid fuels R&D program at all levels: industry, government laboratories, and academia, to expand the knowledge base to enable the development of technologies that can reduce the fuel consumption of transportation systems powered by ICEs. The near-term introduction of such technologies into existing production facilities will reduce the growth in transportation petroleum use during a transition to alternative power plants and power-train configurations. This is the focus of the Partnership's advanced combustion and emission control (ACEC) technical team.

The overarching goals, technical targets, and program structure of the ACEC technical team are basically the same as reported in the Phase 2 review of the program (NRC, 2008). The technical team has established the following technical engine target goals for 2010:

- Engine peak brake thermal efficiency (BTE): 45 percent
- Nitrogen oxides (NO_x) and particulate matter (PM) emissions: Tier 2 Bin 5 (T2B5)
- Power-train cost: <$30/kW

The general focus of the ACEC technical team's work to achieve these targets continues to be lean-burn, direct-injection engines for vehicles fueled by diesel, gasoline, and biofuel or other alternative fuels, provided appropriate carbon emission mitigation is accomplished during their production. Within this broad area specific foci include the following:

---

[2] Of course, this can vary depending on the economic expectations of consumers, who may change their behavior depending on the state of the economy.

[3] In this discussion, hybrid vehicles are included as ICE power trains fueled with a liquid HC fuel. In the hybrid, the energy source is the HC fuel; the hybridization allows more optimal use of the engine and vehicle power-train system.

- Low-temperature combustion (LTC)
  —Control
  —Expanding the load range
  —Coupling to fuel characteristics
  —Transient operation
  —Combustion mode switching
- Aftertreatment
  —Diesel particulate filter (DPF) modeling
  —Lean NO_x traps
  —Selective catalytic reduction (SCR) NO_x reduction
  —Potential catalyst identification for HC NO_x catalysis
- Tool development
  —Improved computational fluid dynamic (CFD) capabilities
  —Improved diagnostics capabilities
  —Comparison of CFD and experiment

In this quest, all aspects of the engine and power train are under investigation. Individual subsystems and processes, such as injection systems, turbochargers, combustion chamber system optimization, the enhanced use of alternative combustion processes (such as low-temperature combustion) and exhaust-gas energy recovery, are actively being investigated. All aspects of the engine operation are being pursued. The electrification of auxiliaries, matching the engine operation to the fuel characteristics, and reducing friction through advanced lubricants are subjects of investigation. Advanced sensors and total power-train system optimization will be enablers for integrating alternative combustion processes into the engine operational map. This will enable optimal matching of the engine and the exhaust aftertreatment systems. In addition, improvements in the aftertreatment systems, particularly lean NO_x systems, will be a critical component of meeting the technical team's targets. Current exhaust-gas aftertreatment systems increase fuel consumption. More effective exhaust emission systems will have a double benefit. They will reduce the fuel consumption associated with their use, and they will allow the engine to be tuned differently with an attendant increase in efficiency.

Hydrogen-fueled ICEs have also been investigated. Such technology could allow a broader use of hydrogen within the transportation system and thus allow the implementation of a hydrogen infrastructure while chemical-electric conversion power plants penetrate the market. However, the hydrogen-fueled ICE vehicle will have similar energy density and specific energy constraints as those of an HFCV, described above.

In all these endeavors, the key hurdle continues to be detailed fundamental understanding of the chemical, thermal, and physical processes taking place within the power train and combustion system.

Good progress is being made by the ACEC technical team in meeting the technical targets. A peak thermal efficiency for an ICE of 43 percent has been achieved. A peak engine efficiency of 45 percent has been achieved for a hydrogen-fueled ICE. The operational range for LTC has been enhanced through active cylinder valve actuation and intake boosting. The technical team reported achieving engine loads of 16 bar (1.6 MPa) indicated mean effective pressure (IMEP) with homogeneous charge compression ignition (HCCI) using a combination of exhaust-gas recirculation (EGR) and intake boost. Additional sensing devices are being developed and integrated into the engine cylinder and power train that facilitate better control of the in-cylinder conditions and power-train energy flow management, which is a necessity for the integration of LTC operation into the engine map.

To maximize the gains in reducing fuel consumption and emissions, every aspect of the ICE power train and aftertreatment system must be optimized for every operating condition in the vehicle's duty cycle. This requires accurate control and manipulation of all engine control parameters for each operating condition. The fundamental research being performed by the ACEC technical team is generating the knowledge base necessary for the identification of how to optimize the combustion process at any operating condition. This understanding is being incorporated into detailed CFD simulations, which in turn accurately replicate the experimental results with minimum adjustable numerical tuning.

The predictive capabilities of the current CFD codes are very good. In fact, the codes are now being used to guide experiments and, more importantly, to identify the combination of engine control parameters that will optimize the engine and power-train performance at different operating conditions, including the use of different combinations of fuels. This is a significant technical accomplishment.

The simulation currently being used is KIVA III, developed by the U.S. Department of Energy (DOE). KIVA is an open-source-code program, which allows researchers to incorporate new understanding directly into the code for any aspect of the thermophysical processes occurring within the engine: for example, improved kinetic schemes for different fuel types, or new submodels that more accurately represent liquid fuel-combustion chamber surface interactions can be implemented into the code and then exercised for more detailed predictions of combustion results. However, KIVA III is more than 10 years old and lacks important, modern numerical technologies such as parallel computing. Having an up-to-date, open-source-code CFD program for researchers to use is a critical aspect of achieving the improvement potential of the ICE and aftertreatment power trains.

To conduct such a program successfully requires close coordination among industry, government laboratories, and academia. The ACEC technical team continues to do a good job with this close coordination. The organizational structure of the team's activities involves memoranda of understanding (MOU) between companies and government laboratories, working group meetings, regular intergroup reviews, and an annual peer-reviewed research meeting. The technical team's responses to the recommendation of the previous review were good (DOE, 2009c).

The energy companies continue to be engaged, and the program of Fuels for Advanced Combustion Engines (FACE), organized under the Coordinating Research Council (CRC), is supplying an important database on the impact of fuel characteristics on engine-emission processes and alternative combustion process facilitation.

The technical team has had difficulty specifically addressing its cost target. The team has assumed that the base engine cost will be $20/kW, and the incremental cost for the technology improvements, which includes enhanced aftertreatment, will be $10/kW. The team has not been able to confirm these estimates with public domain data. Consequently, it has adopted a strategy of determining the technical feasibility of the power train and aftertreatment system, and from there it will work on reducing costs by system improvements (i.e., reduce engine-out emissions, maximizing use of LTC, improving aftertreatment robustness to poisons and thermal degradation, reducing precious metal content).

Funding

The FY 2009 funding level for the ACEC technical team was $25.4 million, with the requested level for FY 2010 being $27 million: the funds appropriated for FY 2010 were $34 million. A breakdown of how the FY 2009 funding was dispersed among different organizations and technologies is shown in Figure 3-1.

Adjustments and New Issues

Since the National Research Council's (NRC's) Phase 2 review of the FreedomCAR and Fuel Partnership research program (NRC, 2008), changes in the country's energy situation have occurred. The biofuels program has grown significantly. Estimates that up to 30 percent of U.S. liquid HC energy could be displaced by domestically produced biofuels have appeared in the literature.[4] A genetically modified alga has attracted attention as a way to enhance the recycling of power plant's CO_2 emissions into a viable transportation fuel.[5] The prospect of enhanced electric storage capacity has spurred the interest in plug-in hybrid electric vehicles (PHEVs). And, new, more stringent emission regulations for NO_x and PM are scheduled to go into effect after 2010. All of these will impact the ACEC program.

The ACEC technical team has acknowledged these changes and addressed them in its future plans. For example, the team is now engaged in fundamental combustion, emission, and kinetic studies of fuel derived from biomass. This work

[4] See for example < http://www.altdotenergy.com/2009/02/sandia-gm-study-finds-large-scale-biofuel-is-sustainable/>.

[5] See for example <http://www.exxonmobil.com/Corporate/energy_climate_con_vehicle_algae.aspx>.

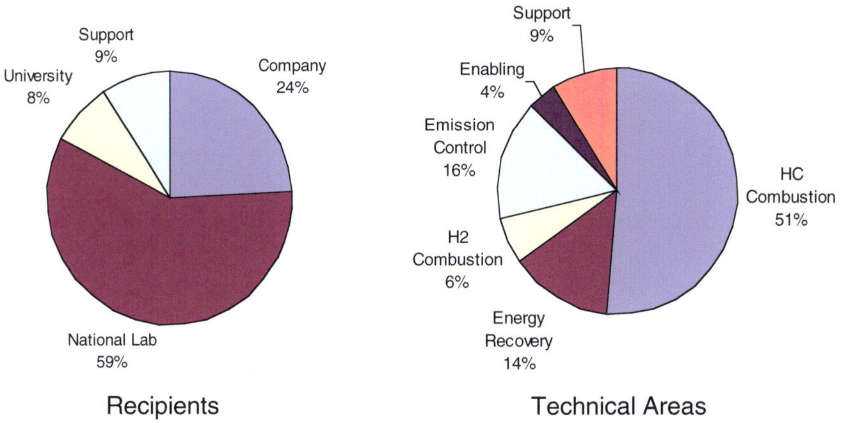

FIGURE 3-1 DOE advanced combustion engine research and development funding, FY 2009. SOURCE: Advanced Combustion and Emission Control Technical Team, Presentation to the committee, August 4, 2009, Southfield, Michigan.

is aimed at understanding the fundamental changes that occur in ignition and emission-formation processes when different compounds, such as methyl esters that are found in biofuels, are used in the engine. The auto-ignition characteristics of many oxygenates, which occur naturally in biofuels, may offer advantages in expanding the range of low-temperature combustion or in expanding the optimal-efficiency regions in engine maps.

The new emission standards will require that the vehicle emission target will need to be changed from Tier 2 Bin 5 to Tier 2 Bin 2. With this change will come new challenges in lean NO_x aftertreatment, specifically mitigating the impact of sulfur poisoning and the associated degradation of the system performance that occurs with repeated desulfurization.

Pending fuel economy standards will impact the vehicle mix as the on-the-road light-duty vehicle fleet turns over. Vehicles will become smaller and lighter. Thus the requirements for the engines and power trains will change. For example, the optimal engine for a PHEV will be significantly smaller than the engines typical in vehicles today. All of these changes will force an evolution in engine and power-train design, and consequently, the optimal power-train configuration, operating scenario, and fuel characteristics will also evolve. It is likely that the operational targets for the engine and power train will become more fluid.

To the committee the foregoing considerations raised the question of whether system-level modeling could be used as a tool to evaluate the optimal power train, engine map, and fuel characteristics for different scenarios of vehicle, power train, and fuel mixes as the energy market and government regulation evolve.

Recommendations

Within the scope of the FreedomCAR and Fuel Partnership objectives, the funding level and work allocation for the continued development of the ICE and vehicle electrification seem appropriate. The ACEC technical team is doing a good job of maintaining a close and constructive working relationship with the stakeholders within the vehicle and energy community. It is critical for the technical team to maintain this collaboration and to look for ways to make it even stronger.

The largest barrier to implementing advanced combustion, aftertreatment, and fuel technologies continues to be an insufficient knowledge base. Not only topic-specific understanding but also an understanding of the system-level interactions among the energy carrier, the energy release process, and the final emission cleanup are critical to continued improvement of the ICE power train.[6] Continued close collaboration between the DOE and industry is necessary to allow newly developed technologies to transition into the industrial laboratories and to lead to the identification of new areas where enhanced understanding will be the most beneficial.

Recommendation 3-1. The DOE should continue to support financially, be active in, and work to further enhance the collaborations among the national laboratories, industry, and academia in order most effectively to direct research efforts to areas where enhanced fundamental understanding is most needed to improve internal combustion engine and aftertreatment power-train performance.

Recommendation 3-2. The DOE should continue to support the development and dissemination of the open-source-code computational fluid dynamics program KIVA. This tool is critical to integrating the new understanding of combustion and emission processes into a framework that allows it to be used to guide further research and identify fuel and engine operating conditions that will maximize reductions in fuel consumption over the entire operating range of the engine.

Recommendation 3-3. The advanced combustion and emission control technical team should engage with the biofuels research community to ensure that the biofuels research which the team is conducting is consistent with and leverages the latest developments in the field of biofuels R&D.

Recommendation 3-4. As the vehicle mix within the on-the-road light-duty vehicle fleet is likely to change with the implementation of the new fuel economy standards, the advanced combustion and emission control technical team should

---

[6] As with the discussion in this section, hybrid and even plug-in hybrid power trains are included in the general classification of power train.

interface with the system modeling technical team to make sure that their research programs are consistent with the changing demands for the optimal matching of the engine operational regimes, power management, and emission control that will be imposed on the internal combustion engine and hybrid power trains as the vehicle characteristics evolve.

FUEL CELL SUBSYSTEM

The fuel cell power-generation subsystem—containing the fuel cell stack and its balance of plant (BoP) consisting of the supporting air and fuel supply, thermal management, and controls—is arguably the most complex and challenging element of the entire hydrogen-fueled vehicle. As this technology is not yet fully developed, advancements are needed to meet the established efficiency, durability, lifetime, and cost targets. Although there are multiple approaches and engineering configurations under development by the original equipment manufacturers (OEMs; the automobile manufacturers), the burden of successfully accomplishing all advancements by any one organization is challenging, since much of the effort is high-risk and demands the assignment of critical resources.

The Department of Energy has been proactive in providing fuel cell R&D support for the precompetitive scientific and engineering initiatives that are high-risk and enabling by providing funding to appropriate organizations such as universities, national laboratories, and the private sector. In many cases involving private-sector developers, R&D activities have the added benefit that the initiatives may lead to supply chain development. Such support has been available through the open solicitation process for nearly 8 years under this current program (FreedomCAR and Fuel Partnership) and a number of years prior in forerunner efforts such as the Partnership for the Next Generation of Vehicles (PNGV). The recent years have witnessed funding activities on fuel cells through multiple DOE organizations, including the Office of Energy Efficiency and Renewable Energy (EERE), Basic Energy Sciences (BES), the Small Business Innovation Research (SBIR) office, and more recently, with coordinated efforts with the National Energy Technology Laboratory (NETL) and the National Renewable Energy Laboratory (NREL). During this period, multiyear development programs have resulted in awards in support of fuel cell R&D efforts. In this program alone, the 8 years of funding has resulted in three cost-shared solicitations, resulting in many R&D contracts ranging from early programmatically focused efforts, to "go/no-go" milestone-based R&D. As a result of these programs, the core technology has advanced in such areas as fuel cell membranes, catalysts, operating modes, durability, lifetime, and the scientific evaluation of the factors limiting performance (e.g., gas quality), to name a few, while projected costs have continually decreased. The activities have been coordinated directly by the fuel cell technical team organized under the FreedomCAR and Fuel Partnership Executive Steering Group (ESG).

With respect to this review, since FY 2007 approximately $140 million (see Figure 3-2) has been appropriated in total to support the attainment of the fuel cell technology roadmap R&D (DOE, 2009a) objectives so that the Partnership's chances of meeting the 2010 targets and the 2015 commercialization-readiness decision goal are enhanced. In order for this decision to be reasonably made (i.e., for the OEMs to decide by 2015 whether or not to initiate the next steps in the process of developing commercially viable vehicles based on a hydrogen fuel cell power-generation subsystem), much of the technology must be demonstrated to be operational in vehicles, or at least it must be significantly beyond laboratory scale. The attainment of, or progress toward, 2010 targets, as shown in Table 3-1 for selected fuel cell stack targets, can also be considered as a measure of progress of the program. The 2010 goals assessment is also a measure of ascertaining whether the R&D topics initially deemed to be the highest priority are still appropriate. In such cases, the DOE go/no-go decision-making process can be and is employed. The committee's assessment is that the fuel cell technical team is well coordinated and is aligned with respect to the achievement of the goals and the longer-term, high-risk technology challenges, especially as the OEMs are now road testing prototype HFCVs.

In light of the prior funding of this program as reported in this review period (2007-2009) and the advancements reported to the committee, at the time of this assessment the success of the program could have been put in jeopardy as a result

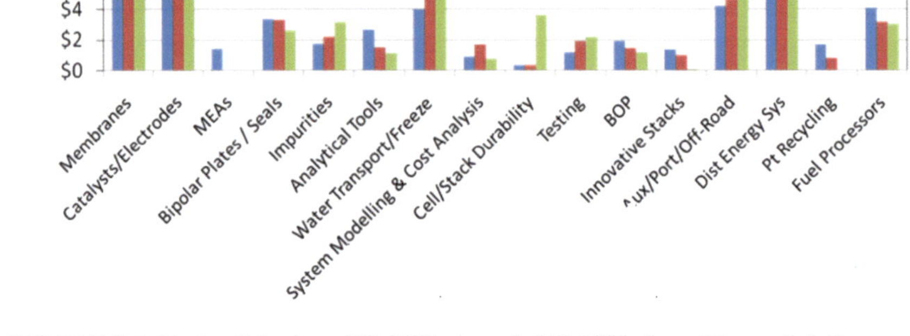

FIGURE 3-2 Fuel cell budget, FY 2007 through FY 2009 (in millions of dollars per year).
SOURCE: C. Gittleman (GM) and K. Epping Martin (DOE), "FreedomCAR Fuel Cell Technical Team," Presentation to the committee, August 4, 2009, Southfield, Michigan.

TABLE 3-1 Selected Fuel Cell Stack Targets and Progress

	Current[a]	2010 Target	2015 Target
Lifetime (hr)	1,977		5,000
Catalyst loading (mg/cm^2)	0.15	0.30	0.15
Efficiency at 25% rated power	59%[b]	60%	60%
Projected system costs (500,000 units produced per year; $/kW)	~60-70	45	30
Power density (W/L) without storage	224		325
Specific power (W/kg) without storage	406		325

[a] As reported to the committee at its August 4-5, 2009, meeting and by S. Satyapal, DOE, Hydrogen Program Overview, Annual Merit Review and Peer Evaluation Meeting, May 18, 2009, Washington, D.C. Available on the Web at <http://www.hydrogen.energy.gov/pdfs/review09/program_overview_2009_amr.pdf>.

[b] Based on laboratory results from 3M and not full-size modules.

of the zeroing of the primary budget line items related to the fuel cell development activities (in the FY 2010 administration's budget request). If vehicle fuel cell development is to continue, such funding must remain intact and must be directed at the R&D that can help enable OEMs to develop the complete vehicle fuel cell power-generation subsystem. More specifically, as stated in its recommendations, the committee believes that technologies needed for vehicle fuel cell systems—and not just fuel cells for stationary, auxiliary power, or portable applications—should be pursued. Vehicle fuel cell requirements can be, and usually are, different and more challenging with respect to cost, reliability, and manufacturability when compared to the other nonvehicle applications. Furthermore, continued funding, especially of the high-risk concepts, will help facilitate next-generation technologies.

The fuel cell stack is composed of layers of catalyzed proton-conducting membranes and electrode assemblies (MEAs) that react supplied hydrogen fuel with oxygen from the air. The MEAs must operate under all environmental conditions and have nearly turnkey operating characteristics. The continued refinement of prior generations of the MEAs is a major issue, as neither the earlier nor current versions have been shown to meet simultaneously the 2015 targets for performance, lifetime, reliability, and cost. However, significant progress has been and continues to be made, as evidenced by field and laboratory testing. Table 3-1 presents selected fuel cell stack targets, the current status, and the progress against such targets as reported by the DOE and the Partnership. Even with such data, complicating the comprehensive understanding of the status of the fuel cell technology is the fact that the OEMs have their own respective (proprietary) fuel cell activities and engineering approaches, which may or may not be synchronized with the DOE-funded development efforts. With that said, what has been reported is that, overall, the OEMs have shown increased power density for the fuel cell stack and BoP, while at the same time the packaging and operating modes have become quite sophisticated. Manufacturing aspects of the power-generation subsystem have yet to become a serious focus partly because of the continuing evolution of the technology (i.e., capital funding for fixed assets is not prudent when the technology may still change). Yet, selected subcomponent suppliers have prototype manufacturing capability today that would meet near-term demand. A noteworthy comment on significant achievements since the previous review is that, while almost every major target has been met in one form or another, they have unfortunately been in separate initiatives and not from a collective, single source. Although it is not definitive that the 2015 targets are achievable by the year 2015, the promising results to date indicate that they could be.

As the DOE programs address precompetitive R&D, it is important to point out again that the OEMs have their own proprietary engineering programs and are not obligated to incorporate DOE-funded developments and technologies into their units. As a result, aside from the open reporting of such performance data and improvements, the contributions of the publicly funded programs and the degree to which the results impact the success of the OEMs related to efficiency, durability, lifetime, and cost are not known with certainty.

Assessment of the Program and Key Achievements

Results reported from the recently funded activities indicate that the current fuel cell subsystem program is making significant progress, yet the successful attainment of the 2015 targets will not be known for some time. However, the attainment of the 2010 targets will be a very positive indicator of future success. Key achievements highlighted by the DOE and made since the Phase 2 review (NRC, 2008) are primarily performance- and cost-related: in particular, fuel cell stack technology tested under realistic on-road operating conditions. Demonstrated stack lifetimes in on-road vehicles have increased from operating times of approximately 1,250 hours to 1,977 hours.[7] With the goal of 5,000 hours, this represents a significant achievement since the Phase 2 NRC review. Furthermore, single-cell and short-stack tests at the laboratory scale have demonstrated (using accelerated test protocols) much longer run times (3M Company, 7,200 hours)[8] that, if demonstrated in vehicles under realistic on-road conditions, would meet or exceed the goals of the Partnership. Larger-scale stack performance and on-road testing will help to validate the laboratory data and determine the ultimate value to the program.

Cost (reduction) is the other area where significant advancements have been reported. The cost assessment of a fuel cell power plant is difficult to make, since the stack and BoP materials and system technology are still evolving.

[7] See DOE's Annual Merit Review on the Web at <http://www.hydrogen.energy.gov/annual_review.html>.

[8] See DOE's 2009 Annual Merit Review, presentation by S. Satyapal, on the Web at <http://www.hydrogen.energy.gov/pdfs/review09/program_overview_2009_amr.pdf>.

Furthermore, such component costs are not benefiting from established volume manufacturing operations at this time. To complicate the assessment of future cost further, the fuel cell stack is dependent on the platinum metals markets and on ever-changing global metals markets dynamics. In making cost projections, the assumptions are many and in some cases are based on still-unproven laboratory-phase performance. Although the results are encouraging, the same conclusion that was reached in the Phase 2 review (NRC, 2008) still holds: the cost projections are highly dependent on many unknowns and must have greater resolution in the forthcoming period. However, two separate DOE-funded studies, with independent oversight, have concluded that at volumes of 500,000 units per year, the cost per kilowatt for the fuel cell subsystem, including the fuel cell and BoP, will be approximately \$60-\$70/kW (Satyapal, 2009; James and Kalinoski, 2009; Sinha et al., 2009). These figures are still over two times higher than the target, but significantly lower than the \$107/kW presented during the Phase 2 review. The projected cost is split nearly evenly between the stack and the BoP. Furthermore, within these cost assessments it was pointed out that platinum and membrane costs are still significant hindrances to stack cost reduction (currently active areas of DOE-funded efforts). Both stack and BoP cost reductions are required in order to achieve the \$30/kW target. It was suggested by the cost studies that system simplification is essential to reduce the BoP cost.

Another measurement of progress is the number of granted patents related to FreedomCAR technology which have been derived from DOE funding. Such a metric is indicative of technology that is in the marketplace today or is available for commercialization. It impacts the fuel cell developers as well as the supply chain. As reported in a Pacific Northwest National Laboratory (PNNL) report prepared for the DOE on the patents originated from the Hydrogen, Fuel Cells and Infrastructure Technologies (HFCIT) program (DOE, 2009d), of the 144 patents, 70 have been issued since 2002 when the FreedomCAR program was initiated. Such patents have been awarded to universities, the private sector, and the national laboratories, and they represent inventions in all segments of the technology.

A particular subcomponent impacting the fuel cell cost and lifetime is the membrane and electrode assembly. Catalyst quantities required to support the hydrogen and oxygen reaction also contribute to both metrics. The lower catalyst loadings, although attractive from a cost perspective, introduce a greater risk of negatively impacting performance. The lower the catalyst loadings, the greater the potential impact on performance. Loadings as low as 0.2 mg/cm^2 have been reported in full-size modules, yet the direct impact on life is not clear at this time. The ability to achieve less than 20 grams of precious metal per 80 kW stack has been verified in the laboratory but has not yet been demonstrated in a vehicle. Progress in other MEA areas has been mixed. Current membranes have a greater degree of robustness but are still impacted by secondary reactions, which can lead to chemical attack and therefore failure. Newer, lower-cost membrane development activities have been funded in recent years, and although the results of such work look promising, it is unclear if they will lead to significant improvements in stability, life, and cost.

Overall, within this last period of activity and considering that the results available for assessments stem from 2001-2008 activities, much progress has been made in key areas. However, the coordination of the program (targets) by the fuel cell technical team could be reevaluated in some areas, such as the following:

- The system being modeled by the Argonne National Laboratory (ANL) and used for costing efforts by the two cost contractors is not equivalent to the system expected to be under test for the 2010 goals assessment or the 2015 commercialization-readiness decision. Even though the DOE uses the system model principally for the purpose of costing and not system performance, the model should be representative of the actual system. As a specific example of the disparity, the costing was performed using a two-stack configuration, although even in the 2010 goals-assessment configuration it is expected that there will be a single stack.
- On-road vehicle operation and performance trials have proven to be invaluable in uncovering unanticipated problems and verifying operation, and yet there is no plan to continue the funding for this activity.
- As the majority of on-road vehicles were tested in moderate climates, additional assessments of performance in all-weather conditions are needed to provide additional insight into the viability of the current technology path.

Significant Barriers and Issues That Need to Be Addressed

The barriers remaining for the fuel cell subsystem R&D program are both programmatic and technical. Programmatic issues relate to the coordination and execution of the high-risk research in order that the solicitation timing and content address updated requirements of the Partnership.

Technical barriers that still remain for the fuel cell stack are membrane and electrode life and cost. Both areas must remain the focus of the next round of solicitations. Further, as indicated in the preceding discussion on cost, system simplification is essential to cost reduction.

Response to Phase 2 Recommendations

The Partnership addressed and concurred with the majority of the recommendations from the National Research Council's Phase 2 review (DOE, 2009c; NRC, 2008). In some instances the FY 2009 DOE budget reflects such recommendations, and the Partnership continued to be proactive in specific areas highlighted in the Phase 2 report. In particular, the focus on advanced membranes and catalysts to address the cost, reliability, and durability challenges is reflected in the current budget.

Appropriate Federal Role

The committee believes that federal funding for fuel cell activities is appropriate and that it remains extremely important, especially for the high-risk-related technical barriers. The need will be reduced, however, as the OEMs move closer to a commercialization phase and as the companies lock in designs for their engineering solutions. New concepts, cost-reduction R&D, and alternative engineering approaches must remain the focus of the DOE funding, especially for the development of next-generation technologies. This is especially important because numerous subsystems are interrelated. Furthermore, supply chain R&D and manufacturing concepts might require funding for the high-risk initiatives.

As the number of potential vehicle fuel cell manufacturers has been reduced in the current (2009) time frame, it is extremely important to maintain continuity and commitment regarding fuel cell technology from the perspective of the United States. As European and Asian car manufacturers are announcing fuel cell vehicle commercialization target dates in the 2015 time frame, the role that the DOE plays in supporting the FreedomCAR and Fuel Partnership has become even more critical.

Conclusions and Observations

Technology has advanced since the NRC Phase 2 review, and it is progressing even in spite of the current economic and automotive industry challenges. Together with the DOE, the OEMs with their proprietary engineering advancements have reported significant on-road achievements toward the 5,000-hour reliability and durability target. Although it is difficult to assess the specific technologies adopted by the OEMs, and the origins of the technologies, the degree of success is apparent.

The core stack technology advancement appears to be one of the most significant achievements reported to date. Although the current approach is very promising, there is a risk that down-selection of any one specific technology might be premature. Backup and secondary approaches must be in place, especially with respect to the high-risk elements.

Results to date indicate that most of the 2010 fuel cell performance targets are going to be met. The attainment of the majority of the 2015 targets is still difficult to predict.

The coordination of activities between the fuel cell technical team members and the DOE appears functional and focused. Yet, because of the nature of the DOE multiyear funded solicitation process and the rapid advancements by the OEMs, there can be a divergence of the currently funded efforts and the fuel cell subsystem R&D needs.

Recommendations

Recommendation 3-5. As the auto companies begin to down-select technologies for fuel cell vehicles, they must focus their limited R&D resources on development engineering for the platform selected and move into the competitive (as distinct from precompetitive) arena. The only way that alternative fuel cell systems and components can receive sufficient attention to mitigate the overall program risk is for the precompetitive program, sponsored largely by the DOE, to support them. Thus, the DOE should increase its focus on precompetitive R&D related to both the fuel cell stack and the balance of plant—the other components of the fuel cell system required for successful operation, such as controls, fuel storage, instrumentation, and so forth—to develop alternatives to the down-selected technologies.

Recommendation 3-6. The DOE should incorporate more of the advanced, most recent, nonproprietary OEM system configuration specifications in the various systems and cost models for fuel cell power plants. Systems configurations no longer demonstrated to be optimal should be abandoned in favor of best proven technology.

Recommendation 3-7. The DOE should establish backup technology paths, in particular for stack operation modes and stack components, with the fuel cell technical team to address the case of current technology selections determined not likely to meet the targets. The DOE should assess which critical technology development efforts are not yielding sufficient progress and ensure that adequate levels of support for alternative pathways are in place.

Recommendation 3-8. The DOE, with input from the fuel cell technical team, should evaluate, and in selected cases accelerate, the timing of the "go/no-go" decisions when it is evident that significant technological progress has been made and adopted by the OEMs.

ONBOARD HYDROGEN STORAGE

Background

Onboard hydrogen storage is a key enabler for fuel-cell-powered vehicles. The primary focus of the hydrogen storage program within the FreedomCAR and Fuel Partnership is to drive the development and demonstration of commercially viable hydrogen storage technologies for transportation and stationary applications. A specific goal of the program is a vehicle driving range of greater than 300 miles between refuelings while simultaneously meeting vehicle packaging, cost, and performance requirements. The program also includes life-cycle issues, energy efficiencies, safety, and the environmental impact of the applied hydrogen storage technologies.

The primary focus of the program is exploratory materials concepts for onboard storage with the potential to meet the long-term goals. Issues for high-pressure tanks that may have nearer-term application and can benefit from exploratory research are also included. Concepts developed in this program could potentially benefit all hydrogen storage applications.

The work of the onboard hydrogen storage program is organized in four centers of excellence (COEs): the Chemical Hydrogen Storage COE, the Hydrogen Sorption COE, the Metal Hydrides COE, and the Hydrogen Storage Engineering COE (see Table 3-2 and Figure 3-3). In 2009, DOE-funded activities in hydrogen storage, including Office of Science Basic Energy Sciences awards, were carried out at 41 universities, 15 companies, and 14 federal laboratories. The hydrogen storage technical team and the DOE provide guidance for the work of the COEs. The program also includes several independent projects that are not associated with any of the COEs (see Figure 3-3). The hydrogen storage technical team is a joint technical team with participants from both the automotive and the fuel industries.

The four COEs and the independent projects constitute the framework of the National Hydrogen Storage Project (see Box 3-1 and Figure 3-3). The independent research projects explore promising hydrogen storage materials and concepts, off-board hydrogen storage for hydrogen delivery, the standardized testing of hydrogen storage properties, and analyses of life-cycle cost, energy efficiency, and environmental impact for hydrogen storage systems.

The EERE hydrogen technology budget appropriation for hydrogen storage was $59.2 million in FY 2009, which was 36 percent above the FY 2008 appropriation ($43.5 million) for applied hydrogen storage research (see Table 3-3). The FY 2010 EERE appropriation for hydrogen storage is $32.0 million. This reduced funding versus FY 2009 will meet existing grant commitments but provides no new starts. The BES budget within the Office of Science also included support of Basic Energy Research Needs for the Hydrogen Economy ($38.3 million in FY 2009). Novel materials for hydrogen storage were a high-priority area for BES funding, receiving $8.0 million in FY 2008 and $9.0 million in FY 2009.

Hydrogen storage has been an R&D priority for the DOE for less than a decade. The committee believes that continued activity with adequate R&D funding should be provided for material-based storage in order to increase the marketability of HFCVs. New focus and funding should be given to compressed-gas storage in order to meet near-term needs for hydrogen storage.

Current Status Vis-à-Vis Goals and System Targets

The physical storage of hydrogen on vehicles as compressed gas (and to a lesser extent liquid hydrogen) has emerged as the technology path for the early introduction of HFCVs. The hydrogen storage capacity of tanks is performance limiting for some vehicle architectures, but hydrogen storage overall is not a

TABLE 3-2 Centers of Excellence (COEs) Project Focus and Participating Organizations

COE	Project Focus	Organizations
Center of Excellence on Chemical Hydrogen Storage	New chemical hydrogen storage materials and regeneration processes, including ammonia borane, ionic liquids, heteroatom containing organics, catalytic processes, and new concepts for hydrogen release and spent-fuel regeneration.	Los Alamos National Laboratory, Pacific Northwest National Laboratory, Intematix Corporation, Millennium Cell, Northern Arizona University, Pennsylvania State University, Rohm and Haas, Inc., University of Alabama, University of California-Davis, University of Missouri, University of Pennsylvania, University of Washington, US Borax
Center of Excellence on Hydrogen Sorption	High surface area sorbents including metal-carbon hybrids, boron-carbon materials, metal organic frameworks, nanohorns and fibers, conducting and porous polymers; modeling and mechanistic understanding	National Renewable Energy Laboratory, Air Products and Chemicals, Inc., California Institute of Technology, Duke University, Lawrence Livermore National Laboratory, National Institute of Standards and Technology, Oak Ridge National Laboratory, Pennsylvania State University, Rice University, University of Michigan, University of North Carolina, University of Pennsylvania
Center of Excellence on Metal Hydrides	Light-weight complex hydrides, destabilized binary hydrides, intermetallic hydrides, modified lithium amides, and other advanced onboard reversible hydrides	Sandia National Laboratories-Livermore, Brookhaven National Laboratory, California Institute of Technology, General Electric, HRL Laboratories, Intematix Corporation, Jet Propulsion Laboratory, National Institute of Standards and Technology, Oak Ridge National Laboratory, Savannah River National Laboratory, Stanford University, University of Hawaii, University of Illinois at Urbana-Champaign, University of Nevada-Reno, University of Pittsburgh/Carnegie Mellon University, University of Utah
Hydrogen Storage Engineering Center of Excellence	Energy challenges associated with developing low-pressure material-based hydrogen storage systems for enabling onboard storage of hydrogen for fuel-cell-powered vehicles and for achieving customer expected driving range and performance. (Includes systems integration, prototype development, and systems analysis.)	Savannah River National Laboratory, Pacific Northwest National Laboratory, United Technologies Research Center, Los Alamos National Laboratory, NASA Jet Propulsion Laboratory, National Renewable Energy Laboratory, General Motors Company, Ford Motor Company, Oregon State University, Lincoln Composites, Inc.

SOURCE: DOE (2009a), Section 3.3, Hydrogen Storage.

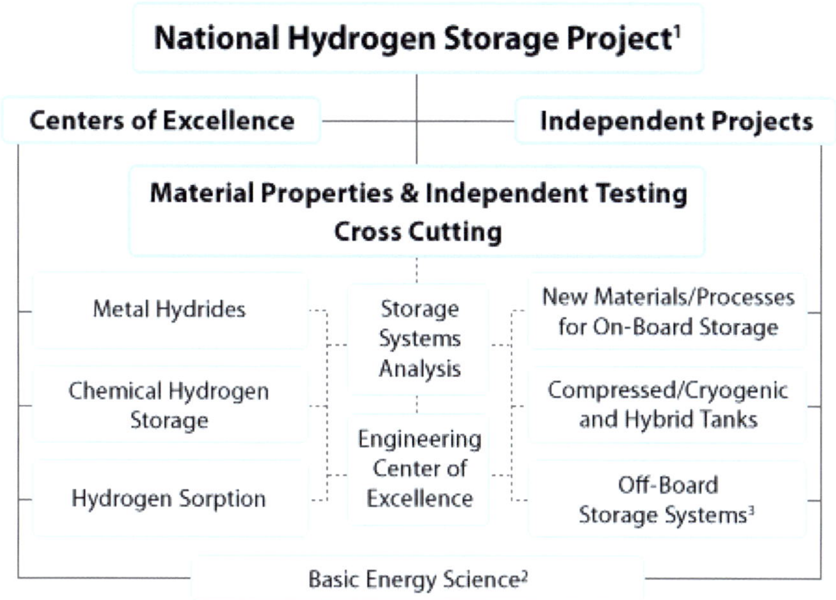

1. Coordinated by DOE Energy Efficiency and Renewable Energy, Office of Fuel Cell Technologies.
2. Basic science for hydrogen storage conducted through DOE Office of Science, Basic Energy Sciences.
3. Coordinated with Delivery Program element.

FIGURE 3-3 Structure of the National Hydrogen Storage Project. SOURCE: Reprinted from DOE (2009a).

BOX 3-1 Fiscal Year 2009 Participating Organizations: Independent Projects in Hydrogen Storage

Industry	Universities and Institutes	Federal Laboratories
Air Products and Chemicals, Inc.; Gas Technology Institute; H2 Technology Consulting LLC; Quantum Technologies; TIAX; UOP; UTRC	Alfred U.; Hydrogen Education Foundation; Michigan Tech; Missouri-Columbia; Northwestern, Penn State; Purdue; Southwest Research Institute; SUNY-Syracuse; U of Arkansas; UC Berkeley; UCLA; UC Santa Barbara; University of Connecticut; U Penn/Drexel	ANL; SRNL; LANL; LLNL; ORNL; SNL

SOURCE: Adapted from Satyapal (2009), p. 33.

TABLE 3-3 Office of Energy Efficiency and Renewable Energy Budget Appropriations for Hydrogen Storage, FY 2007 through FY 2010 (millions of dollars)

Fiscal Year	Appropriation ($ millions)
2007	33.7
2008	43.5
2009	59.2
2010	32.0

SOURCE: A. Sudik, F. Bavarian, and N. Stetson, Hydrogen Storage Joint Technical Team, Presentation to the committee, August 5, 2009, Southfield, Michigan.

"blocking" technology for vehicle introduction. The storage capacity of current tanks does not meet the long-term goals, but it may be adequate for some applications for which the cost can be justified. Thus, research aimed at significantly higher hydrogen storage capability needs to be kept as a research objective. Materials-based storage at the level required to meet all program targets is considered theoretically achievable, yet no material has been identified that meets all of the targets. These results are promising but will not be achieved without adequate funding, which is required to continue to make progress and to attract outstanding scientists and engineers to this line of research. All targets (weight, volume, efficiency, cost, packaging, safety, refueling ability, etc.) must be met simultaneously. The discovery and development of materials for onboard hydrogen storage remain high-technical-risk R&D in need of research attention and government funding.

The targets and timing for the onboard hydrogen storage program were revised since the Phase 2 review to reflect the knowledge gained from real-world vehicle experience and the vehicle weight and space appropriate for market penetration. The revised targets assume that the vehicle architecture will change between gasoline ICE and HFCVs. The newly revised hydrogen storage targets are shown in Table 3A-1 in the annex at the end of this chapter.

The overall objective for hydrogen storage remains unchanged except for the targets: vehicle performance across vehicle models with acceptable driving range, packaging, and cost, while meeting all safety requirements. Hydrogen storage capacity and cost are key parameters for initial materials evaluation. The revised targets are as follows:

- By 2010, develop and verify onboard hydrogen storage systems achieving (old targets) 2 kWh/kg (6 weight percent [wt%]), 1.5 kWh/L, and $4/kWh; (new targets) 1.5 kWh/kg (4.5 wt%), 0.9 kWh/L (28 g/L).
- By 2015, develop and verify onboard hydrogen storage systems achieving (old targets) 3 kWh/kg (9 wt%), 2.7 kWh/L, and $2/kWh; (new

targets) 1.8 kWh/kg (5.5 wt%), 1.3 kWh/L (40 g/L). (See "Annex to Onboard Hydrogen Storage" at the end of this chapter.)

Since the Phase 2 review, more than 350 materials approaches for hydrogen storage were investigated, of which 68 percent have been discontinued and 32 percent are still under investigation. Twenty-one hydrogen storage patents were issued. To date no material for onboard hydrogen storage has been identified that meets the full set of 2015 targets. These system targets are listed in the annex to this chapter.

Milestones achieved since the Phase 2 NRC (2008) review include the following:

- The no-go decision made for vehicle hydrogen storage during the Phase 2 review was to discontinue applied R&D in pure, undoped, single-walled carbon nanotubes based on the fact that they were not able to meet the storage target of 6 wt% close to room temperature (2006).
- A no-go decision was made for sodium borohydride onboard vehicular hydrogen storage (2007).
- The Multiyear Research, Development, and Demonstration Plan was developed for the years 2005-2015 (2007).
- A Hydrogen Storage Engineering Center funding opportunity was announced (2008).
- The down-select decision on chemical hydrogen storage materials was made (2008). Selection criteria were established (e.g., gravimetric capacity, potential to regenerate onboard, regenerable, acceptable phase change, H_2 release rate materials stability, endothermic release, H_2 release temperature). Of 120 materials and classes of materials examined to date, 15 percent were selected for continued study.
- Metal hydrides materials were down-selected. Selection criteria were established based on the potential to meet 2010 technical targets. Of 74 materials investigated to date, 40 have been selected for further work.
- The Hydrogen Sorption COE has investigated 160 materials, and 35 percent are still in its inventory. A down-select report is in preparation.
- The announcement of the Hydrogen Storage Engineering COE (2009) was made. This COE will address system integration and prototype development in coordination with the materials centers. It was awarded to the Savannah River National Laboratory (SRNL).
- An H-Prize competition notice was issued for "Breakthrough Advances in Materials for Hydrogen Storage" (DOE, 2009b). A single amount of $1 million will be awarded for the development of an onboard hydrogen storage material that meets or exceeds a set of performance targets specified in the competition announcement. This prize creates an incentive for the R&D community outside the conventional grant process.

- A DOE hydrogen program solicitation was issued for R&D for onboard vehicular hydrogen storage to support the COE or as independent projects (2008).

Assessment of Progress and Key Achievements

The current status of promising hydrogen storage materials is shown in the composite Figure 3-4. Hydrogen storage is shown together with the new targets (volumetric and gravimetric capacity only). Figure 3-4 shows results for the three groups of hydrogen storage materials: complex hydrides, chemical hydrides, and carbon sorbents. Data given here show that all three material groups fall short of the 2015 system targets for both volumetric and gravimetric capacities, but with best results demonstrated for the carbon sorbents.

The current candidate storage materials under investigation for the three classes of materials—reversible metal hydrides, chemical storage materials, and hydrogen sorbents—are listed in Box 3-2.

Information for physical storage is shown in Figure 3-4 for both ambient and cryo-based systems. Data are shown for 350 and 700 bar (ca. 35 MPa and 70 MPa) compressed hydrogen. In the nearer term, ambient physical storage provides a means for advancing the integrated hydrogen fuel cell system development and gaining experience while the materials storage approach is developed further. The ambient systems (the current and simplest configuration) are targeted for the early introduction of the vehicle test fleets.

Expensive "aerospace quality" carbon fiber is needed in the construction of the onboard pressure vessel for hydrogen for HFCVs. Such fibers provide the necessary high strength and lightweight characteristics. The DOE currently has several efforts to reduce the cost of carbon-fiber pressure vessels. Included in these efforts is the use of melt spinning in place of the currently used solution spinning of the PAN (polyacrylonitrile) feedstock. Another project that has promise for cost reduction is the hot-melt processing of PAN. Quantum Technologies is also being funded to reduce the cost of compressed storage by manufacturing process optimization. Also, a carbon-fiber pilot line facility is being funded with American Recovery and Reinvestment Act (ARRA) of 2009 funds at the Oak Ridge National Laboratory to lower the processing and feedstock costs for aerospace-quality fibers.

Two reports, one on cryo-compressed hydrogen (ANL, 2009) and one on compressed hydrogen (TIAX, 2009), released in late 2009, project that the cryo-compressed tank as modeled will meet the gravimetric targets for hydrogen storage but not the volumetric targets. For the compressed hydrogen study, the 350 and 700 bar (ca. 35 MPa and 70 MPa) tanks as modeled will meet the gravimetric targets but none of the volumetric and cost targets. These projections include the balance of plant.

Although the storage density is a critical parameter, all of the targets (weight, volume, efficiency, cost, packaging, safety, refueling ability and time, etc.) must

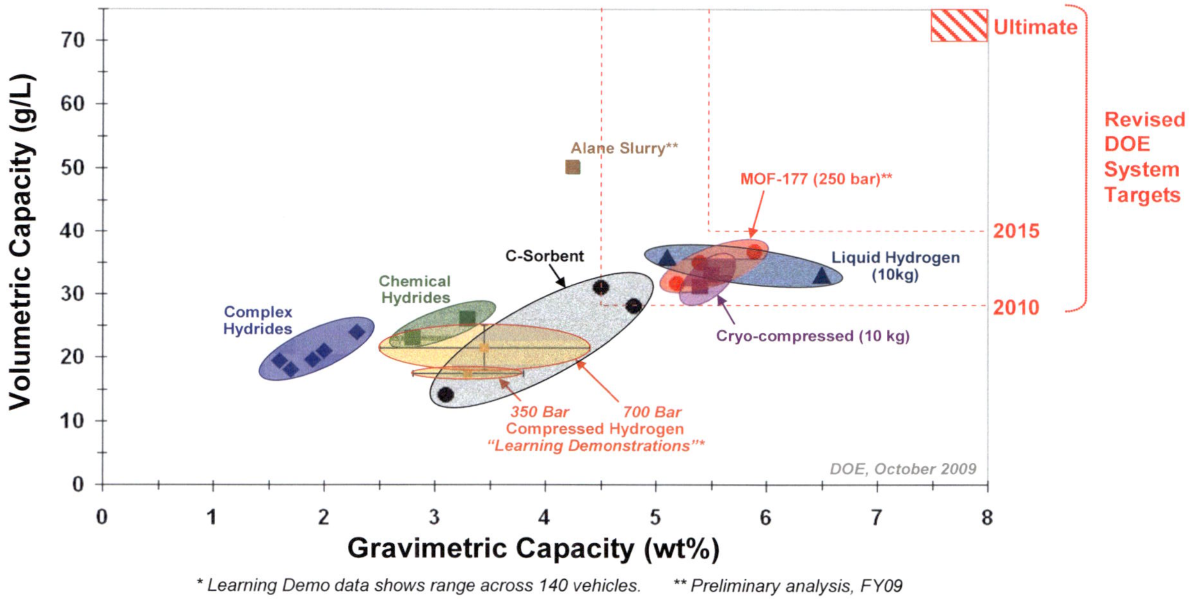

FIGURE 3-4 Current hydrogen storage system status versus revised targets. SOURCE: N. Stetson, DOE, "H_2 Centers of Excellence," Presentation to the committee, October 26, 2009, Washington, D.C.

NOTE: Data based on R&D projections and independent analysis (FY05-FY09) to be periodically updated.

BOX 3-2 Current Candidate Hydrogen Storage Materials Under Investigation

Reversible Metal Hydrides	**Chemical Storage Materials**	**Hydrogen Sorbents**
$Mg(BH_4)_2$	NH_3BH_3 (solid)	Metal doped carbon nanostructures
$Mg(BH_4)_2(NH_3)_2$	NH_3BH_3 (liquid)	Metal organic frameworks
$LiBH_4$	AlH_3	Zeolitic immidozolate frameworks
$LiBH_4/MgH_2$	DADB	Polyether ether ketone derived
$LiBH_4/Mg_2NiH_4$	C-B-N heterocycles	Microporous materials
AlB_4H_{11}	Metal amidoboranes	Covalent organic frameworks
LiMgN	$Al(NH_2BH_3)_3$	Carbide-derived carbon microporous materials
		Spillover materials
		Nanostructured polymeric materials

NOTE: The above are *examples* of some of the current materials and/or types of materials under investigation within the DOE hydrogen storage program portfolio of projects. At this time, no material has been found that meets the requirements for gravimetric and volumetric capacities, hydrogen release and uptake rates at acceptable temperatures and pressures, cycle life, impurity tolerance and release, and costs.
SOURCE: Communication to the committee from the Office of Energy Efficiency and Renewable Energy, December 2009.

be met simultaneously. None of the approaches (neither material-based nor physical storage) meets the combined targets. The program approach of using the Hydrogen Storage Engineering COE to fabricate and evaluate complete vehicle-ready test systems is an excellent technique for selecting the most viable material configuration. The material and physical storage results to date (obtained in a short time) as well as the Hydrogen Storage Engineering COE are promising with respect to the attainment of the 2015 objectives.

Also of note, the DOE Vehicle/Infrastructure Demonstration Program reported having achieved an HFCV range of 196 to 254 miles. The highest HFCV range reported to date (estimated to be 431 miles on a single full tank of compressed gas from Toyota) is the result of a field evaluation for a fuel cell hybrid vehicle. This field test included data analysis by NREL and SRNL through a collaborative research and development agreement (CRADA).

Highlights of Technical Accomplishments

Several significant technical accomplishments have been achieved, as follows:

- Overall progress in system capacity is reported to have increased 50 percent since 2007.
- Systems analysis of hydrogen storage options has been accomplished (Argonne National Laboratory).

- Areas have been identified for materials-based and physical/compressed storage-system cost reduction (TIAX LLC).
- $MB_{12}H_{12}$ effects on borohydride reversibility have been studied (NASA Jet Propulsion Laboratory [JPL], California Institute of Technology, and General Electric).
- Alane regeneration has been achieved by means of adduct (Brookhaven National Laboratory).
- Ammonia borane regeneration efficiency and yields have improved (COE on Chemical Hydrogen Storage).
- Hydrogen binding energy on adsorbents has been increased (University of California, Berkeley; University of California, Santa Barbara; and Texas A&M).
- Improved hydride kinetics by means of carbon aerogel scaffolds has been achieved (HRL Laboratories, LLC; and Lawrence Livermore National Laboratory).
- A subscale prototype has been developed for $NaAlH_4$.
- A full-scale prototype has been developed for cryo-compressed hydrogen storage.

Significant Barriers and Issues That Need to Be Addressed

The hydrogen storage program has good recognition of the many technical needs and challenges that it faces. The following is a compilation of these issues:

- Those common to all storage approaches
  —*System weight and volume:* Too high for meeting the 300-mile range across a wide spectrum of vehicle platforms. Basically no suitable storage material has been identified and developed.
  —*System cost:* Needs to be reduced compared with petroleum. Cost areas include materials of construction and manufacturing methods, and balance of plant components.
  —*Charging time (refueling) for material storage:* Storage capacity and rates of sorption and release need understanding and improvement (goal is 3 minutes for 5 kg charge).
  —*Energy efficiency:* Charging and discharging of hydrogen to a storage tank or storage material can be an energy consumer (requiring heating and/or cooling), which impacts the overall system efficiency.
  —*Systems issues:* Thermal management, durability and operability, hydrogen quality, containment vessels, dispensing technologies, and system life-cycle assessment and prediction need to be addressed.
  —*Codes and standards:* Needed for entire system and for all interfaces.
  —*Safety:* Issues related to hydrogen storage (see Recommendations 2-2 and 3-10).

- Reversible materials-based systems (reversible onboard)
  —*Hydrogen sorption (physisorption and chemisorption) and desorption processes:* Understanding of these processes is needed.
  —*Reproducibility of performance:* Needs to be demonstrated.
- Chemical hydrogen storage systems (typically regenerated off-board)
  —*Regeneration process:* Process cost, efficiency, environmental impact.
  —*By-product/spent material removal:* Important issues to be addressed.
- Pressurized hydrogen storage tanks
  —*The cost of high-quality carbon fibers:* Needs to be reduced.

Technical tasks have been established that address each of these issues.

Future Plans

The newly organized Hydrogen Storage Engineering COE has taken on the coordination of the engineering aspects of material-based hydrogen storage systems. This center plans crosscutting hydrogen storage activities organized across six areas: performance analysis, system modeling, enabling technologies, materials operating requirements, transport phenomenon, and subscale prototype construction and testing and evaluation. Given the fact that the completion dates for the other three COEs and a number of independent projects fall within FY 2010, this COE will have a critical role in capturing the progress for a sustained activity during any transition period.

Target dates have been appropriately set for technology down-select decisions:

- A complete analysis of onboard storage options for 2010 and 2015 targets was scheduled for 2009 as well as a decision point on advanced carbon-based materials and a down-select for chemical hydrogen storage approaches for the 2010 targets.
- A decision on reversible metal hydride R&D is scheduled to be made in the fourth quarter of 2010 as well as a decision point on chemical hydrogen storage R&D.
- The down-select for onboard reversible hydrogen storage materials and for chemical hydrogen storage approaches with the potential to meet 2015 targets is set for the fourth quarter of 2013.
- Complete laboratory-scale prototype system and evaluation against 2015 targets is scheduled for the fourth quarter of 2015.

Future plans include continued R&D on breakthrough hydrogen storage materials with increased emphasis on engineering analysis, a broadening of the effort to include all of the targets (versus just the capacity targets), and increased coordination between the basic and applied activities. Early market applications will receive increased emphasis. The hydrogen storage technical team will continue to monitor and leverage globally activities on hydrogen storage. In order to address the fuel storage needs of and to set priorities for fuel cell applications, the EERE plans to conduct a Request for Information (RFI) and a workshop during FY 2010.

Response to Recommendations from the Phase 2 Review

The DOE agreed with most of the recommendations from the Phase 2 review (DOE, 2009c; NRC, 2008). It did not address in the response the safety implications of relying on compressed-gas storage in the interim period. Compressed-gas tank safety needs further attention. The FY 2009 budget appropriation allowed the program to be supported at a high level for continuing and new R&D activities. The program has been managed to balance resources—for example, to cut storage approaches without potential and to down-select approaches with the most potential. The Hydrogen Storage Engineering COE is a timely use of resources and fits well with the other COEs. The real-world experience with pressurized tanks is providing information on R&D issues. The DOE (2009c) stated in the response that the issue of materials for pressurized tanks is being addressed in other parts of the program and in future solicitations. In response to the recommendation for a strong basic research portfolio, it was noted that BES held a contractors' meeting for principal investigators funded on projects related to the Partnership in conjunction with the DOE Hydrogen Program Annual Merit Review and Peer Evaluation Meeting in 2006 and again in 2009. Hydrogen basic research is well funded in the FY 2009 program, and new concepts will continue to be supported.

Appropriate Federal Role

The federal sponsorship of the hydrogen storage activities within the FreedomCAR and Fuel Partnership is an appropriate federal role. The research work supported is high-risk and potentially important for meeting national energy and emission objectives. This sponsorship has significantly stimulated research and aided the advancement of the field through its support of a significant number of qualified researchers, providing focus on common goals, maintaining communications among participants, and peer review of results.

Recommendations

Recommendation 3-9. The centers of excellence are well managed and have provided an excellent approach for organizing and managing a large, diverse research activity with many participants at various locations. Measures should be taken to continue research on the most promising approaches for onboard hydrogen storage materials. The complete documentation and communication of findings should be undertaken for all materials examined for the completed R&D. Furthermore, in view of the fact that the hydrogen storage program has been in place for less than a decade, the Partnership should strongly support continuing the funding of basic research activities. Public domain contractor reports should be available through links on the DOE EERE Web site.

Recommendation 3-10. Research on compressed-gas storage should be expanded to include safety-related activities that determine cost and/or weight, such as validation of the design point for burst pressure ratio at beginning of life and end of life and evaluation of Type 3 versus Type 4 storage vessels. Furthermore, finite-element modeling of stresses and heat flow in fires, investigative work on wraps (i.e., translation efficiency), and analysis of applicability of compressed-gas storage to specific vehicle types would be beneficial.

Recommendation 3-11. The high cost of aerospace-quality carbon fiber is a major impediment to achieving cost-effective compressed-hydrogen storage. The reduction of fiber cost and the use of alternative fibers should be a major focus for the future. Systems analysis methodology should be applied to needed critical cost reductions.

Recommendation 3-12. The hydrogen storage program is one of the most critical parts of the hydrogen/fuel cell vehicle part of the FreedomCAR and Fuel Partnership—both for physical (compressed gas) and for materials storage. It should continue to be funded, especially the systems-level work in the Hydrogen Storage Engineering COE. Efforts should also be directed to compressed-gas storage to help achieve weight and cost reduction while maintaining safety.

Recommendation 3-13. The time for charging the hydrogen storage material with hydrogen (refueling time) is a program goal (3 minutes for a 5 kg charge). Concepts beyond materials properties alone should be explored to meet this challenge for customer satisfaction, and will require coordination with the areas of production, off-board storage, and dispensing.

Recommendation 3-14. There should be an effort to anticipate hydrogen storage material property and performance requirements that will place demands on developed systems—for example, purity and response to impurities, aging and lifetime prediction, and safety in adverse environments. Linkage between the hydrogen storage and production and delivery activities should receive attention.

Recommendation 3-15. The search for suitable onboard hydrogen storage materials has been broadly based, and significant progress is reported. Nonetheless the current materials are not close to the long-range goals of the Partnership. Onboard hydrogen storage R&D risks losing out to near-term applications for future emphasis and funding. The management of a long-term/short-term joint portfolio should be given consideration.

ELECTROCHEMICAL ENERGY STORAGE

Introduction

Electrochemical energy storage technologies, batteries, and ultracapacitors are critical to the advancement of the FreedomCAR and Fuel Partnership's long-term goals. Significant improvement in their performance can result in battery electric vehicles (BEVs), one of the ways to meet the Partnership's goal of "energy freedom, environmental freedom, and vehicle freedom." The FreedomCAR and Vehicle Technologies (FCVT) program (now renamed the Vehicle Technologies [VT] program), has supported the advancement of batteries and ultracapacitors from the beginning as a key to developing hybrid electric vehicles (HEVs). Also, before a hydrogen fuel infrastructure is fully developed, plug-in hybrid electric vehicle (PHEV) and BEV technologies, which would compete with hydrogen fuel cell vehicles, may offer a transitional means to improve fuel efficiency and emissions reduction. Since the success of HFCVs is not assured, this transition role could turn out in many cases to be a more permanent scenario.

In 2006, in response to the President's Advanced Energy Initiative, the FCVT program began the development of PHEVs, or extended-range electric vehicles. In contrast to conventional vehicles or HEVs, PHEVs are able to drive on electric power alone for some distance, depending on the electric battery storage capacity. PHEVs thus need more advanced batteries and electric power components than HEVs need. Electric energy storage technologies have taken on an even greater importance in the past year due to the priorities of the new administration to "put 1 million plug-in hybrid cars—cars that can get up to 150 miles per gallon—on the road by 2015, cars that we will work to make sure are built here in America."[9] Furthermore, corporate average fuel economy (CAFE) standards were increased 40 percent to a national fuel economy standard of 35 miles per gallon (mpg) by 2020 (the Obama administration is targeting 2016 rather than 2020). This increase provides a regulatory incentive to increased HEV and PHEV production.

In 1999, HEVs were first introduced in the United States, and their market penetration continued to grow through 2007. In 2008, their sales decreased with the general decrease in all auto sales. Overall the number of HEVs sold has increased 271 percent from 2004 to 2008—from 84,000 to 312,000 vehicles—yet this represents only about 2.5 percent of the new vehicles sold in 2008. The number of models available has also increased from 5 in 2004 to 18 in 2008. All of the HEVs available use a nickel metal-hydride (NiMH) battery, and the DOE has been involved in the advancement of this technology since the 1990s. However, the

[9] See, for example, <http://www.whitehouse.gov/agenda/energy_and_environment/>.

NiMH battery will not meet the long-term FreedomCAR electrochemical energy storage goals for HEVs of a 15-year life with 25 kW pulse power and a cost of $500 by 2010. Thus, the Partnership, through the VT program, is focused on the development of lithium-ion (Li-ion) batteries for HEVs. Major improvements of Li-ion technology are one key requirement for the economic mass production of competitive PHEVs, HFCVs, and BEVs. Li-ion-powered BEVs began production in 2008 with the introduction of the Tesla Roadster powered by 6,800 cells sized for commercial electronics. (Tesla is an expensive sports car that does not meet the target goals of the Partnership.) In addition, a large number of auto companies have announced their intention to launch HEVs, PHEVs, and BEVs using Li-ion batteries in the next few years.

The VT program, in collaboration with the United States Advanced Battery Consortium (USABC), manages the electrochemical energy storage technology program with a goal of the advancement of battery technologies, to the point that the program partners are encouraged to introduce hybrid and electric vehicles with large market potential. Technology development is undertaken by battery manufacturers, DOE national laboratories, and universities, and by awards through the SBIR program. The effort is composed of three subactivities: (1) Battery Technology Development is involved in battery system module development, including design and fabrication specifications, testing procedures, cost modeling and recycling studies, and technology assessment and the benchmark testing of various battery systems; (2) Applied Battery Research focuses primarily on improving the understanding of failure and life-limiting parameters, including safety and abuse tolerance, of the Li-ion system that currently is closest to meeting the technical goals; and (3) Long-Term Battery Research addresses the fundamental understanding of specific electrochemical systems for Li-ion batteries and the development of newer couples with a potential for higher power and energy density.

The Partnership's budget for electrochemical energy technologies has increased as the importance of PHEV battery development has increased. The budget was increased from $24.4 million in FY 2006 to $40.8 million in FY 2007, with a significant increase primarily for PHEV batteries. It was again increased to $48 million in FY 2008 and to $69 million for FY 2009. The FY 2009 budget included $15 million for HEV systems, $38 million for PHEV systems, and $16 million for exploratory R&D. The budget request for FY 2010 is for $78 million. Also, full battery system development is done in collaboration with the USABC through competitive subcontracts that are at least 50 percent cost-shared.

In addition, the Advanced Research Projects Agency-Energy (ARPA-E) of the DOE continues to fund several projects on energy storage technologies for both stationary and vehicular applications. The focus of these projects is primarily to develop high-energy-density batteries. Furthermore, fundamental research projects on electrochemical energy systems are funded by the BES. The VT program contributes about $2 million to the BES for this effort. The BES focuses on long-term needs, such as a basic understanding of materials, interfacial charge transfer, and the development of tools and processes for the design of new materials. Although the BES mandate on energy storage is broader and longer term, it works in close coordination with the VT program to advance the energy storage needs for automotive applications.

Of the 312,000 HEVs sold in 2008, only 31,000 (10 percent of total HEV sales) were manufactured by the three U.S. auto companies, whereas Toyota Prius sales comprised about half of the total sales of HEVs. In order to accelerate the manufacture and deployment of electric vehicles, batteries, and related power components here in America and to create thousands of jobs in these technologies, 48 new advanced battery and electric drive projects of $2.4 billion were funded under the ARRA. Of these funds, $1.5 billion in grants is for producing batteries and their components and expanding battery-recycling capacity. Although the ARRA funding is short term for the purpose of establishing a manufacturing base and primarily increasing employment, it has the potential of influencing continued research and development of advanced batteries into the future.

Until 2007, the FCVT program was primarily involved in the development of high-power electrochemical energy storage systems for HEVs. Since 2007, the FCVT program has expanded the electrochemical energy storage activity to include PHEVs. The goal is to develop vehicles that would allow a 40+ mile electric range, enough to satisfy about 70 percent of the daily commuting travel in the United States. These vehicles operate in both modes—electric-only (as in a BEV) and electrical/mechanical (as in an HEV)—and the battery can be recharged from a standard electric outlet. The VT efforts for PHEVs are directed at developing higher-energy batteries that meet the targets (see Table 3-4) established by the DOE and USABC for commercial viability. In addition, it continues to pursue research activities toward even-higher-energy batteries for BEV applications.

Program Status and Assessment

Lithium-ion battery technologies hold promise of achieving the long-term goals of high power, energy, and other performance requirements for HEV and PHEV applications at lower anticipated costs than those for other battery systems. Thus, the Partnership is correctly focused on the development of these technologies while it continues to benchmark competing battery technologies and encourages research on higher-energy chemistries for BEV applications.

Three Li-ion battery chemistries classified by the cathode material, including (1) lithium nickel, cobalt, and aluminum; (2) lithium iron phosphate; and (3) lithium manganese spinel and a carbon anode have been developed and tested for HEV applications. Sufficient progress has been made on these chemistries that they meet or exceed most of the 2010 performance goals listed in Table 3-4. Since the Phase 2 review, there has been improvement in discharge and regen-

TABLE 3-4 Target Characteristics for Hybrid Electric Vehicle Batteries for 2010

Characteristics	Unit	Status in 2009	Minimum Goal	Maximum Goal
10 s discharge pulse power	kW	29.5	25	40
10 s regenerative pulse power	kW	35.3	20	35
Available energy	Wh	780	300	500
Efficiency	%	>90	90	90
Cycle life	Cycles	200,000	300,000	300,000
Calendar life	Years	15	15	15
System cost at 100,000/yr	$	1,035	500	800
Maximum system weight	kg	36.5	40	60
Maximum system volume	Liter	35	32	45
Maximum operating voltage	V	140	≤400	≤400
Self-discharge	Wh/day	<50	50	50
Cold cranking power at –30°C	kW	6	5	7
Operating temperature range	°C	+10 to +35	–30 to +52	–30 to +52

SOURCE: Available at <http://www1.eere.energy.gov/vehiclesandfuels/pdfs/mypp/3-2_hybr_elec_prop.pdf>.

erative pulse power rating, calendar life as measured by accelerated testing, and increased cycle life. There is still room for improvement in the operating temperature range and cold-cranking capability. The projected cost of the battery is still very high, about twice the target of $500 when produced in quantities of 100,000 units per year. Work continues in order to increase performance, reduce cost, and improve the safety of these batteries. The most notable achievement of the Li-ion battery development program for HEVs has been the announcement by Mercedes and BMW that they will use Li-ion batteries in their next generation of hybrid cars.[10]

These three Li-ion chemistries, which are the most advanced for HEV applications, are also being developed for PHEV applications. The PHEV allows for flexibility in the energy being used to power the wheels, whether it is electricity from batteries or fuel powering the ICE. Thus, it allows for flexibility and complexity in the power architecture design of the PHEV power train. As discussed in further detail in the section below on "Electric Propulsion and Electrical Systems," a series drivetrain powers the vehicle only by an electric motor using electricity from the battery. The battery is charged from the electricity grid or by the vehicle's gasoline engine by means of a generator. Such a design is being considered by General Motors for the Chevy Volt. In a parallel drivetrain, there is a direct connection between the engine and the wheels. Therefore, the vehicle

[10] See, for example, <http://www.hybridcars.com/news/mercedes-lithium-ion-hybrid-2009.html>.

TABLE 3-5 Target Characteristics for the Years 2012 and 2014 for Plug-in Hybrid Electric Vehicle Batteries

Characteristics at End of Life	Unit	High Power/ Energy Ratio, 2012	High Energy/ Power Ratio, 2014
Equivalent electric range	miles	10	40
Energy for charge depletion (BEV mode), 10 kW rate	kWh	3.4	11.6
Energy for charge sustaining (HEV mode)	kWh	0.5	0.3
10 s discharge pulse power	kW	45	38
10 s regenerative pulse power	kW	30	25
Efficiency	%	90	90
CD cycle life	Cycles	5000	5000
CS cycle life (50 Wh)	Cycles	300,000	300,000
Calendar life	Years	10	10
System cost at 100,000/yr	$	1700	3400
Maximum system weight	kg	60	120
Maximum system volume	Liter	40	80
Maximum operating voltage	V	≤400	≤400
Self discharge	Wh/day	50	50
Cold cranking power at 30°C	kW	7	7
Operating temperature range	°C	–30 to +52	–30 to +52

SOURCE: Adapted from Howell (2009).

can be powered by electricity and the gasoline-fueled engine simultaneously, or by the gasoline-fueled engine only. Such a design is being considered by Toyota in a plug-in version of the Prius. In this design the mechanical and electrical power are blended, and the degree and criteria for blending can be varied. The PHEV architecture plays an important role in the design of the battery and how it stores energy from the grid, the gasoline engine, or from regeneration during braking. In BEV applications the vehicles run on electricity only, and thus high-energy-density batteries are required. In HEV applications the vehicle runs primarily on gasoline, and thus high-power batteries are required, but in PHEVs the batteries may require high energy or high power depending on the architecture design of the drive train and the range sought. The two cases of high power-to-energy ratio and the high energy-to-power ratio battery characteristics for PHEV applications are listed in Table 3-5. Further details on energy storage and power electronics are contained in the PHEV R&D plan.[11]

The design of a PHEV battery requires the simultaneous optimization of power, energy, and life while maintaining safety and reducing cost. There are

[11] See <http://www1.eere.energy.gov/vehiclesandfuels/pdfs/program/plug-in_summary_rpt.pdf> and <http://www1.eere.energy.gov/vehiclesandfuels/pdfs/program/phev_rd_plan_june_2007.pdf>.

inherent trade-offs among the various requirements. Generally, increasing the energy density will decrease the power density, whereas increasing the power density means using thinner electrodes, which will increase cost, reduce life, and may impact safety. There are also differences in the inherent characteristics among the chemistries. Of the three chemistries, the lithium manganese spinel has the highest power rating, due to its high voltage. The lithium nickelate system has the highest energy density, and the lithium iron phosphate is considered inherently safer than the other two systems. Thus, the Partnership has followed multiple paths of development using different materials and designs to optimize performance, life, and cost. At present none of the battery chemistries meets the performance, life, or cost goals for 2012 requirements.

Although the Partnership has not set explicit objectives for battery safety, it clearly is a key element of vehicle safety and its definition by the industry. Some of the battery goals are driven by safety considerations—for example, the requirement of a substantial temperature "window" for the safe operation of cells and batteries. The Partnership has established a series of "abuse" tests to characterize the behavior and safety potential of cells and entire batteries under off-design conditions that might be encountered in practical operation. These include mechanical (crushing, nail penetration, shock), electrical (external shorting, overcharging, and over-discharging) and thermal abuse (heating to above-design temperatures with external and internal sources). The three Li-ion chemistries were tested at both the cell and the battery-pack level in an attempt to access their readiness for use and to improve their design and manufacturability. In addition, significant work was undertaken to obtain a basic understanding of the thermal response of the battery in both normal and abuse conditions to make sure that a condition of thermal runaway does not occur. Because of thermal runaway observed in a substantial number of Li-ion batteries in consumer devices, such as cellular telephones and laptop computers, there are public concerns about the safety of Li-ion batteries in general. However, there are important differences between consumer and automotive battery efforts and applications. In automotive applications, smarter battery-management systems are used; they continuously monitor the battery at the cell level and make corrective action as required. Battery safety thus in large measure is a system characteristic that needs to be managed carefully. Also, the chemistries currently being considered for commercialization in HEVs, PHEVs, and BEVs are different and inherently safer than the $LiCoO_2$ cathode used in consumer applications. The R&D program continues to look for materials that will inherently improve the safety of the system. For example, nano-titanium oxide ($LiTi_{12}O_5$) is being actively investigated as an alternate anode material to replace carbon in order to address the issue of metallic lithium deposition on the carbon anodes in Li-ion cells. Thus the development of the electrochemical couple of lithium manganese oxide spinel cathode and nano-titanium oxide anode is being driven by safety considerations. This electrochemical couple also allows fast charging but at a reduced energy density because of a relatively low voltage.

Although significant progress has been recorded in the Li-ion battery performance, durability, and safety, there has been no improvement in the cost of battery.[12] The projected cost at 100,000 units per year for the HEV application remains at higher than $900, almost twice the 2010 target of $500.[13] It should be noted that as volume builds up the costs are likely to come down. Battery cost will play an even bigger role in the eventual success of the PHEV application because much larger batteries are required. Not only is the size different, but the operating regime is different. A typical HEV battery needs to deliver power to accelerate the vehicle as well as to accept power during regenerative braking. However, the amount of energy storage required is limited to about 10 percent of energy storage on the vehicle. As a result, HEV batteries operate over a limited state of charge (SOC), which enables the battery to deliver many thousands of charge-discharge cycles. In a PHEV application the available energy, which is proportional to the all-electric range of the vehicle, is a more important requirement than the power is. Thus the battery is much bigger and operates over a larger variation of the SOC. The battery may use up to 70 percent of the total energy; however, some manufacturers may limit the used energy to a narrower range to increase life, minimize warranty concerns, and make allowances for the performance (capacity and/or power) deterioration over the life of the battery. Thus in defining the cost per kilowatt-hour of a PHEV, the range of SOC variation and expected life performing at the vehicle requirements need to be specified.

A PHEV battery cost assessment was conducted by TIAX for the DOE (Sinha et al., 2009). The company considered four chemistries (lithium nickel cobalt aluminum, lithium nickel cobalt manganese, lithium manganese spinel, and lithium iron phosphate cathodes, all with carbon anodes), 16 different scenarios (varying electrode loading and percent capacity fade to end of life), and a useful state of charge from 10 to 90 percent. The cost was estimated for a 5.5 kWh usable energy battery (~20 mile electric range) constructed with cylindrical cells only, at a production volume of 500,000 units a year. A sensitivity analysis was conducted for each scenario that estimated the mean cost of the battery to be approximately $360/kWh, varying from $264/kWh to $710/kWh. This results in a cost of $1,450 to $3,900 for a 5.5 kWh battery. Furthermore, the TIAX cost assessment finds that the cost of cathode active material plays a smaller role in the system cost than do cell design parameters, such as electrode loading and thickness, performance factors such as percent fading to end of life, and manufacturing process speeds. TIAX also conducted several "what if" scenarios to determine which variable could reduce the battery cost to $250/kWh (the long-term goal). It was unable to

---

[12] There are issues related to using either battery cost ($) or specific battery cost ($/kWh). The committee has used both, trying to use the most appropriate choice depending on the context and discussion.

[13] Howell, D., and K. Snyder, "Electrochemical Energy Storage," Presentation to the committee, August 4, 2009.

reach \$250/kWh under any of the scenarios considered. It should be noted that the DOE has a target of \$300/kWh for a PHEV-40 in 2014.[14]

The Partnership is commended for conducting this cost study, and it is hoped that such investigations will continue for different conditions and scenarios. The study clearly shows that the cost goals established are very aggressive, and it may be difficult to achieve them using the present chemistries. Thus the DOE should continue its strong support for exploratory research on the fundamentals of electrochemistry and energy storage materials.

At present about 75 percent of the electrochemistry R&D funding is directed to near- and midterm development efforts directed at HEV and PHEV applications and only 25 percent to long-term R&D. The past efforts on HEV and PHEV batteries have borne fruit, and one is now beginning to see application of U.S.-developed technology in prototypical and early commercial HEVs and PHEVs. The Partnership should now take the initiative to strengthen its focus on longer-term research toward high-energy batteries and establishing a path toward BEVs.

The energy storage targets for BEVs were established more than 10 years ago, and the Partnership and the VT program should revisit and update the goals and targets for this automotive segment in view of both the changing market and technology. Several automobile companies have announced the intention to launch BEVs over the next few years, particularly with about 100 miles of driving range for city driving and for fleet usage. There have also been significant increases in R&D activities globally in recent years on novel energy storage materials and systems with promising results. It is imperative that the Partnership increase its effort to maintain the U.S. competitive position. These increased efforts will require increased funding for high-energy batteries and include leveraging all other efforts on electrochemistry and energy storage materials efforts within the DOE and the larger electrochemistry community.

The increasing market share of HEVs and the introduction of PHEVs will result in increasing numbers of advanced batteries in automotive applications. The DOE should initiate a program to develop and pilot the recycling of lithium batteries. Mass adoption of lithium resources would place pressure on global supply, and recycling is an important strategy to mitigate resource depletion and provide an economical supply of the material. It is worth noting that the economics and resource characteristics of battery recycling are driven by the total material content. For example, the economics of recycling current Li-ion batteries is driven by the value of the cobalt contained in the battery (see, e.g., Anderson and Wade, 2001; Xu et al., 2008). Public acceptance will demand stringent health, environmental, and safety standards, especially since one of the main reasons for hybrid vehicles is environmental. The recycling of advanced automotive batteries should be easier than that for small consumer batteries since there are existing programs on the

[14] D. Howell, DOE, "PHEV Update," Presentation to the committee, December 10, 2009, Washington, D.C.

recycling of automotive lead-acid (PbA) batteries. The DOE was correct in providing $9.5 million for the hydrothermal recycling of Li-ion batteries through the ARRA program. The Partnership and the VT program should now follow up by initiating a research program on improved processes for reducing cost and recovering useful materials from this effort. They should also conduct a study to determine the cost of recycling and the potential of savings from recycled materials.

Recommendations

Recommendation 3-16. The Partnership should revisit and modify, as necessary, the goals and targets for battery electric vehicles in view of the changing market conditions and improvements in technologies.

Recommendation 3-17. The Partnership should significantly intensify its efforts to develop improved materials and systems for high-energy batteries for both plug-in electric vehicles and battery electric vehicles.

Recommendation 3-18. The Partnership should conduct a study to determine the cost of recycling batteries and the potential of savings from recycled materials. A research program on improved processes for recycling advanced batteries should be initiated in order to reduce the cost of the processes and recover useful materials and to reduce potentially hazardous toxic waste and, if necessary, to explore and develop new processes that preserve and recycle a much larger portion of the battery values.

ELECTRIC PROPULSION AND ELECTRICAL SYSTEMS

Introduction and Background

The electric propulsion system consisting of power electronics (combinations of a bi-directional dc (direct current)-dc converter, boost converter, and inverter) and one or more electrical machines is needed for HEVs, PHEVs, HFCVs, and BEVs, to provide traction to the wheels from the prime mover. The prime mover for the propulsion system can be an engine, engine-driven generator, battery, or fuel cell, depending on the energy source. In all of these cases, the systems used can be distinguished by the architecture as well as by the size and power. In the subsection below, Figures 3-6 through 3-10 show the major different configurations that apply to each. A vehicle needs other electrical systems such as chargers for electrochemical storage (battery), dc-to-dc converters for the utilities, power management, and a compressor drive for the fuel cell blower; these are discussed below separately.

The Partnership has appropriately focused on key technical areas that are precompetitive with the objective of long-term reductions in size (volume and weight) and cost. To accomplish its objectives, emphasis has been on better packaging, cooling, materials, and devices.

The subprogram Hybrid Electric Systems has a budget for FY 2010 of $146 million within the Office of Vehicle Technologies program budget ($141 million for FreedomCAR and $4.8 million for 21st Century Truck Partnership [21CTP]) and has the following components:

- *Vehicle and Systems Simulation and Testing:* $43.7 million (includes the 21CTP portion),
- *Energy Storage R&D:* $76.27 million, and
- *Advanced Power Electronics and Electric Machines R&D:* $22.29 million.[15]

This section of the report deals with the activities associated with the last item, Advanced Power Electronics and Electric Machines R&D, with an FY 2010 budget appropriation of $22.29 million. The FY 2009 budget was divided as follows: 37 percent for power electronics and 21 percent each for traction drive system, electric machines, and thermal management. The vehicle propulsion system activities are focused on attaining specific hybrid vehicle traction drive performance targets (see Figure 3-5) over the next 10 years for cost, gravimetric and volumetric density, and efficiency through advancements in materials, system design, and component technology. Those advancements would be beneficial and could be applied to any of the four alternative traction systems. These are ambitious goals and perhaps may not be attained in the time frame shown. Meeting these, however, is not as critical for the success of electric propulsion as is meeting the goals for HFCVs (hydrogen storage and fuel cell stack) as well as the battery for BEVs.

At this time it is difficult to predict which type of vehicle (e.g., internal combustion engine vehicle, HEV, PHEV, HFCV, or BEV) will dominate the market in future years. However, it is safe to say that even though the ICE will probably continue to have a large share of the market in the near term, some form of electric propulsion will likely be important in the future. In view of this, the committee believes that additional resources in this area are justified.

Current Status and Assessment

The FreedomCAR and Fuel Partnership focuses on electric drives that require a source of power that provides direct current at voltages of the order of 200 to 450 V. As shown in Figures 3-6 through 3-10, the vehicle power source is a fuel cell, an engine-driven generator, or a battery. Conversion of this power to mechanical power to drive the wheels requires power electronics and one or more

[15] Budget information provided to the committee by Christy Cooper, DOE, January 13, 2010.

Requirements: 55 kW peak for 18 sec; 30 kW continuous; 15-year life; coolant (105°C or air)

Technology Targets

	Traction Drive System				Power Electronics			Motors		
Year	($/kW)	(kW/kg)	(kW/l)	Efficiency	($/kW)	(kW/kg)	(kW/l)	($/kW)	(kW/kg)	(kW/l)
2010	19	1.06	2.6	>90%	7.9	10.8	8.7	11.1	1.2	3.7
2015	12	1.2	3.5	>93%	5	12	12	7	1.3	5
2020	8	1.4	4	>94%	3.3	14.1	13.4	4.7	1.6	5.7

FIGURE 3-5 Hybrid vehicle traction drive performance targets. SOURCE: Rogers (2009).

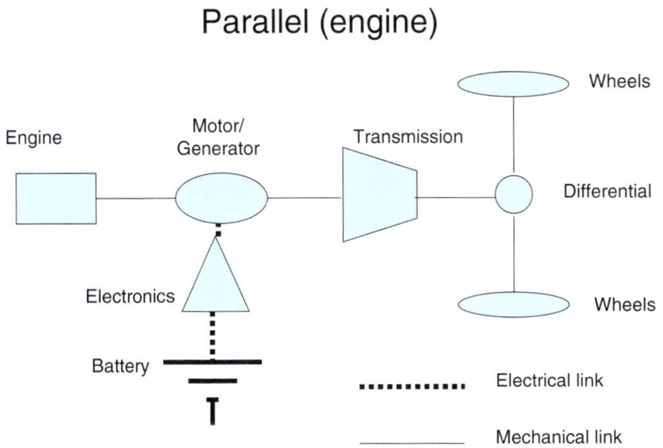

FIGURE 3-6 Schematic of parallel drive configuration for a hybrid vehicle (similar in concept to the Honda Insight Mercedes S series).

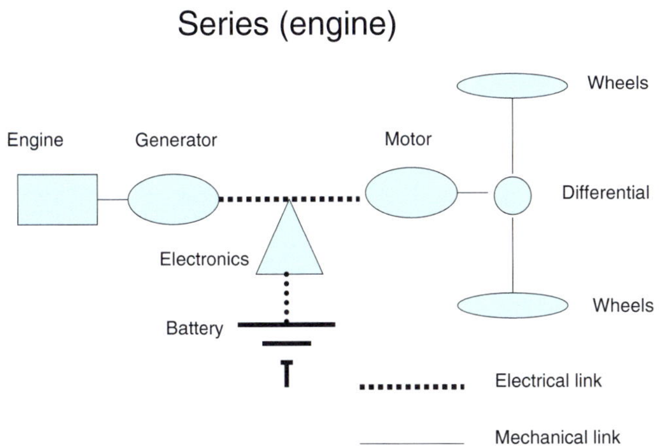

FIGURE 3-7 Schematic of series drive configuration for a plug-in hybrid electric vehicle (similar to the GM Volt).

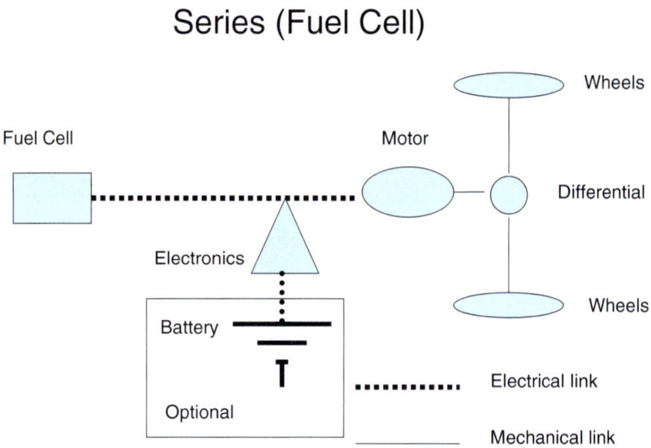

FIGURE 3-8 Schematic of series drive configuration, typical fuel cell vehicle configurations.

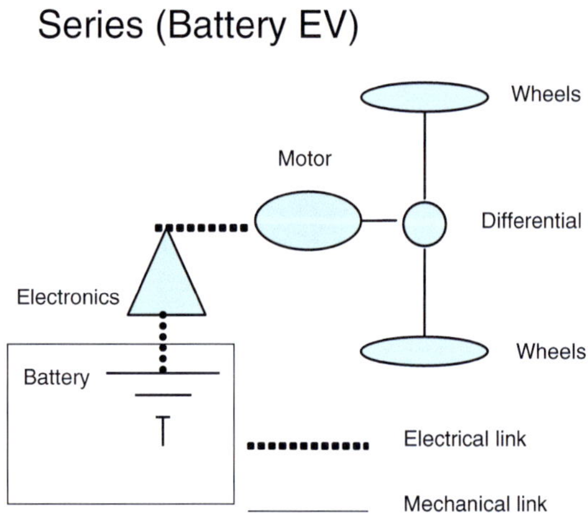

FIGURE 3-9 Schematic of series drive configuration, battery electric vehicle (EV) (similar to the Nissan Leaf and others).

electric motors. Although dc brush motors can be used, this discussion is limited to alternating current (ac) motors (permanent magnet brushless or induction motors) because of their superior performance. Power electronics convert the dc from the source into an ac of variable voltage and variable frequency needed by the motors. It should be noted that these drives have been used for a variety of applications from steel mills to locomotives to appliances, and a great deal of development has taken place. The use of electric propulsion places increased emphasis on

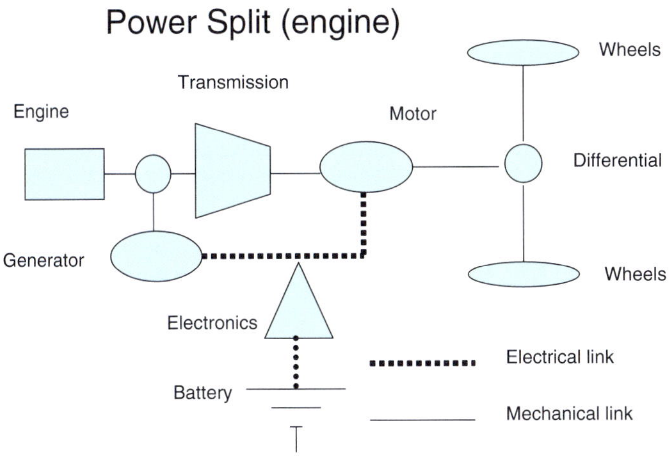

FIGURE 3-10 Schematic of typical power-split hybrid or plug-in hybrid electric vehicle power-train configuration (such as Prius, Escape, and others).

(1) efficiency, to maximize electric range; (2) volume, so that the system can be packaged without reducing space for passengers and cargo; and, of course, (3) cost. Compact and efficient motors and power electronics are essential to all four types of vehicles that the Partnership is working on, namely, HFCVs, HEVs, PHEVs, and BEVs. The present discussion focuses on a review of the traction drive technology status and development efforts to optimize its components for vehicle propulsion, dealing separately with power electronics, electrical machines, and electrical systems.

Power Electronics

The power electronics are composed of a set of semiconductor switches arranged in a block called an inverter, as it converts the dc to ac. Several topologies exist based on control strategies and on whether the frequency conversion is done by the same switches as those for the voltage control. These topologies have been thoroughly investigated over the past 50 years, and basically the selection depends on optimizing the operation. Simply stated, the objectives in the Partnership are to reduce the losses, size, and cost.

Inverter Topology. The inverter changes a dc voltage that varies over narrow limits depending on power to an ac voltage of variable amplitude and frequency depending on motor speed and load; thus the two functions can be performed in two stages (making variable "chopped" dc voltage and then variable frequency) called modulator and inverter or in a single stage called a modulating inverter. It appears that the program has focused exclusively on the modulating inverter topology and may be missing the advantages of separating the two functions.

Separately, efforts are underway by a Delphi-led team[16] to develop a scalable inverter that is capable of being easily sized for a particular application. This approach would dramatically accelerate component and system development time and reduce development costs for all four system types.

Device Cooling. Doped silicon (Si) devices are universally used, and the junction temperature needs to be kept below 125°C. Due to its high band gap and operating temperatures that exceed 250°C, silicon carbide (SiC) offers power inverter efficiencies over silicon. The limitation currently is cost. As noted below, the program is investigating SiC diodes in combination with silicon substrates, and this work needs to continue in spite of today's higher costs. Recently Denso has exhibited SiC-based "power devices."[17]

Switching Speed. The faster the operation the greater the efficiency. Again, devices other than doped silicon, such as the SiC discussed above, have the advantage, and development work should continue.

Components. As stated above for power devices, cooling at higher temperature over ambient is more effective. Power electronics also require capacitors and solders, and in some cases their temperature limits the operation of power electronics. As the cooling of the power devices improves, new materials are needed for both capacitors and solders so that the inverter can operate at a higher temperature.

Thrust Areas in Power Electronics

Scalable Inverter. The concept for a scalable inverter is that it can easily be scaled to meet different power levels. The work on a scalable inverter is a contract of \$8.2 million (\$4.952 million provided by the DOE and \$3.258 million by the contractor) that runs from October 2007 through March 2011 (Taylor, 2009).

SiC Devices. As mentioned in the preceding discussion, SiC devices are better than devices based on doped silicon, because they operate at higher temperatures and have faster switching times. Potentially they lead to smaller and more efficient power electronics. The Partnership is investigating a new process for making SiC

[16] Members and their responsibilities are as follows: Dow Corning/GeneSiC: SiC-on-Si power semiconductor devices; GE: film capacitors; Argonne National Laboratory (ANL): film-on-foil capacitors; Oak Ridge National Laboratory (ORNL): system modeling and simulation, power device characterization, system testing; National Renewable Energy Laboratory (NREL): thermal modeling.

[17] Detroit Auto Show, Cobo Hall, Detroit, January 2010.

on silicon substrates. Building these devices on Si is a desirable first step, since expensive SiC wafers are not used, and should be encouraged.

High-Temperature Capacitors. Developing capacitors that can operate at high temperatures could increase the cooling efficiency and thus reduce the size of power electronics. The following activities are being undertaken:

- *ANL:* This activity uses metal (copper or nickel) foil coated with thin film Pb-La-Zr-Ti-oxide (PLZT) dielectrics. Demonstrated film-on-foil dielectrics with k (relative static permittivity) greater than 1,300, breakdown field greater than 6 MV/cm. Cost projections are not currently available and would greatly depend on the process steps for producing a capacitor, which are still under investigation. An industry manufacturer has not been identified yet.
- *Pennsylvania State University:* This activity uses a flat-panel display glass as a dielectric material and aluminum electrodes. Demonstrated a dielectric constant of 6.2 with a breakdown field of 10 MV/cm. Cost projections are not currently available; laboratory-scale samples are currently expensive, but there is promise in the expanding volume of applications that use flat-panel display glass material. An industry manufacturer has not been identified yet.
- *Sandia National Laboratories (SNL):* This activity uses a high-temperature polymer. The measured dielectric constant is 4.6-4.9 with a breakdown voltage of 1.5 MV/cm. The cost of the dielectric material is low (close to that for polypropylene) for laboratory-scale quantities, and the committee expects that it would be even less expensive for large-quantity production. SNL is currently working with Electronic Concepts, Inc., to produce films at a larger scale.

Packaging and Integration. There are activities ongoing on packaging and integration that include the following:

- Delphi has worked with preferred suppliers to deliver improved silicon integrated gate bipolar transistors and diodes for Delphi's novel packaging solution.
- Delphi and ORNL have investigated many thermal management concepts that have been evaluated and analyzed; several invention records have been written for submission and patent applications.

Silicon-on-Insulator Gate Drivers. The silicon-on-insulator (SOI) project is producing a gate driver circuit to function at temperatures of 200°C (the project is ongoing at ORNL, and hardware exists). It does not focus on any power devices—that is, switches or diodes. The committee believes that sufficient funding currently exists in SiC and gallium nitride (GaN) development elsewhere. It should be noted that the cost is two to four times that of silicon. However, realizing the need for high-temperature drivers to accompany the emergence of high-temperature power devices, this project is an enabler for the higher-temperature operation of inverters and converters. It should be noted that Honda has teamed with Rohm and Haas to develop inverters using SiC devices because of the increase in efficiency as well as the reduction in size because of easier cooling that can be attained in inverter applications. Vehicle implementation is pending safety and cost analysis.[18]

Progress seems to be as follows:

- The SOI gate driver was packaged for high-temperature application using solders. No issues have been found in performance testing of the gate driver at temperature.
- Telefunken has been identified as the fabrication shop. (Telefunken bought Atmel.)

Electrical Machines

In all of the electric drive vehicles, HEVs, PHEVs, BEVs, and HFCVs, an electric motor provides the traction to the wheels, but in some configurations an electrical generator is also needed (see Figures 3-6, 3-7, and 3-10). Primary areas for development are similar to those for power electronics: to reduce size, losses, and cost. The machines are basically of two types and have the following advantages and disadvantages:

- *Permanent magnet brushless motors.* These motors became feasible with the invention of high-energy magnets in the 1980s and currently are used in all electric and hybrid vehicles in production (the only exception being the Tesla Roadster). They have high efficiency, which is critical for vehicles, but the magnets are costly, and they require more complex inverters, as operation in what is known as field weakening mode is limited. To overcome some of these limitations, a configuration known as interior permanent magnet (IPM) designs has evolved. IPMs are used both as motors and generators. The presence of permanent magnets may result in a catastrophic failure if, during driving, there is a short circuit of the winding or a failure of insulation. Since the machine is connected to the wheels, it will continue generating voltage, which will result in an abrupt increase in braking torque as well as possible fire, because much energy is continuously dumped into the short circuit.
- *Induction motors.* These motors are the workhorses in almost all industrial applications. Although less efficient than motors currently used in

[18] See, for example, <http://japancorp.net/Article.Asp?Art_ID=19769>.

> vehicles, they cost less and offer field weakening over a wider speed range. Some original equipment manufacturers (OEMs) have talked about revisiting the choice of motors, and this may be possible if battery costs come down and the premium on efficiency becomes less important than motor costs. Although induction motors are usually used as motors, they can also function as generators. This is obviously important because capturing energy dissipated in the brakes through regenerative braking is essential for the efficient use of the prime-mover energy.

Onboard battery charging during regenerative braking affects the efficiency and cost of the motor. As new battery and motor materials are developed, use of the Powertrain Systems Analysis Toolkit (PSAT), developed at the Argonne National Laboratory under DOE sponsorship, may help quantify material cost and performance trade-offs between motor efficiencies and battery-charging requirements.

Soft Magnetic Materials. The objective in designing new magnetic materials is to reduce two sources of loss, known as (1) hysteresis and (2) eddy current. Conventionally this is accomplished by using thin laminations of steel that contains silicon. The punching and assembly of laminations is expensive, and for years the "holy grail" of soft magnetic materials has been to discover a new material that has both high electrical resistivity and high permeability at the flux density levels needed. As discussed below, the FreedomCAR and Fuel Partnership gave a contract to investigate such materials to General Electric, but it appears that the program was discontinued. Developments in this area, such as the soft magnetic material that Toyota uses in the boost converter in its power electronics, should be monitored (Nozawa et al., 2009).

Cooling. Both liquid and air cooling are conventionally used in vehicles. In the case of hybrid electric vehicles, both oil and engine coolants are available.

Improved Windings. Minimization of the length of winding end turns that does not contribute to output is often used to improve efficiency. Furthermore, an illustration of the lengths to which General Motors and Honda have gone to reduce losses is the fact that they have rectangular conductors, which allow better fill of the slots and thus reduce resistance and improve efficiency.

Thrust Areas for Electrical Machines

A High-Performance Interior Permanent Magnet Machine for Hybrid Vehicles. GE Global Research is the lead organization for a team[19] developing an IPM machine for hybrid vehicles. This effort is a contract of $5.8 million ($3.629 million

[19] The team includes members McCleer Power and the University of Wisconsin-Madison.

provided by the DOE and $2.171 million provided by contractors) that runs from October 2007 through June 2011.

The objective is to build a better permanent magnet motor that is designed to provide 30 kW continuous (55 kW peak) power with a top speed of 14,000 revolutions per minute (rpm), a constant power speed range of 5:1, and an efficiency greater than 95 percent at 20 percent torque (El-Rafaie and Johnson, 2009). Other key objectives are for scalable motors to meet the very tough performance specifications and the use of novel soft magnetic material with a tripling of resistivity enhancement. This is desirable because it improves efficiency. This program is half completed but has the following accomplishments:

- Design of a 30 kW continuous (55 kW peak) motor with a top speed of 14,000 rpm, a constant power speed range of 5:1, and an efficiency greater than 95 percent at 20 percent torque. The latest data show the following accomplishments:
  —*Motor design.* Two rotor and two stator concepts were developed and analyzed in detail. The machine was ready for testing by the end of March 2009.
  —*Low-loss soft magnetic materials.* Bulk amorphous alloy composition was identified and kilogram-scale production was accomplished by gas atomization. A novel microstructure was developed to enhance resistivity and magnetic properties. A composite soft magnetic material with a doubling of resistivity enhancement was demonstrated. However, the results showed that the material had too low a flux density and was prohibitively more expensive (four times the cost of silicon steel).[20]
  —*Low-loss permanent magnet materials.* Hydrogen-based route for processing the high-energy-density magnet materials used in electrical machines. The project has demonstrated a novel composite microstructure to minimize eddy current losses. A permanent magnet microstructure with three to four times resistivity enhancement was demonstrated. This seems very promising, but the committee's information is as of May 2009. Additional information available January 2010 indicated the following:
    - o *Soft magnetic materials:* Some of the work was concluded as not being promising. However, one of the industry awards, General Electric, is continuing the work in the hope of a breakthrough. Clearly, if successful, this would revolutionize the electric motor industry.
    - o *Permanent magnets:* The emphasis will be more on molded high-strength magnets. Although these have lower performance than

[20] Information provided by the DOE to the committee, November 23, 2009.

sintered magnets, they are much easier to manufacture into the complex shapes needed for brushless dc motors.

Electrical Systems

Conventional vehicles have a large number of electrical systems to control emissions, passenger comfort, and safety that are not discussed here. The focus is on the two subsystems—battery chargers and system controllers—used in hybrid, electric, or fuel cell vehicles.

Battery Chargers. Current HEVs use NiMH or PbA batteries that place minimal requirements on the charger. Battery resistance and temperature can be used to derive a reasonable approximation of the SOC during battery operation. This will change with Li-ion, the likely battery of choice in the future. There are several chemistries in use for Li-ion batteries, and they all have the potential of destructive and hazardous "thermal" events if care is not taken during charging and discharging. Such an event resulted in the recall of millions of laptop computer batteries made by Sony in the 1990s. Even though this recall was attributed to a manufacturing defect, the charging voltage of each cell for current chemistries must be controlled to within a few tens of millivolts per cell, and it is expected that the charging voltage of each cell needs to be monitored and that circuits need to be provided to maintain the voltage within safe limits.

There appears to be relatively little work in the Partnership on battery charging. Although it may be argued that this is postcompetitive activity, in the committee's view some work needs to be done to ensure safety and to explore rapid charging. The high voltages used in HEVs require many more Li-ion battery cells in series than is typical in smaller electronic equipment for which chargers are commonly used today. In regard to safety, the number of HEVs on the road is not yet sufficient to evaluate statistically the safety of high-voltage HEV battery packs. Moreover, a reduction in the number of cells in an Li-ion battery pack will likely be required in order to meet the target battery cost. One way that this can be achieved is to increase the cell size and decrease or eliminate the number of cells in parallel. However, this approach will affect safety because heat transfer, end-of-charge control, and cell balancing are all more difficult in larger cells. More work will be required to assess the safety of battery chargers as a function of the cell sizes and battery pack configurations, as well as any changes in the battery chemistry that are ongoing in Li-ion battery development. A three-dimensional performance model of large-format cells may be useful in predicting the over-voltage and temperature variations in large-format cells during high-rate charging (see, e.g., the NREL model, Kim and Smith [2008]). In regard to rapid charging, this strategy is seen as potentially essential to the broad penetration of BEVs. Systematic study of rapid charging implications for battery life and safety of various Li-ion batteries could prove to be of high value in moving forward.

System Controllers. System controllers need to control the vehicle in response to the driver's commands. During acceleration an HEV or BEV should have the "feel" of conventional vehicles, although the high torque produced by electric motors, especially at low speed, is probably an advantage. The controller will also need to control regenerative braking to minimize energy drain from the battery or the fuel cell. It should be pointed out that there is little activity in this area of electrical systems in the Partnership, and this is justified in the committee's view because technology development is in the "competitive areas" of each OEM. The exception to this is a remarkable drive for the compressor expander motor (CEM) of the air supply system for a fuel cell balance of plant as discussed below.[21]

Compressor Expander Motor for Fuel Cell Vehicles. The CEM project incorporates a high-speed drive (165,000 rpm) with a fairly low projected cost. At a production of 500,000 annual units, the projected cost of the CEM is $293 and the cost of the controller is $303, with a total cost of $705 including assembly ($31) and a markup (15 percent for the CEM and 10 percent for the controller). Information provided subsequent to the meeting shows that this is a system similar to a 100,000 rpm demonstration unit from Honeywell. Such speeds are certainly unusual in automotive electric motor applications but, if successful, are very useful for keeping the weight and volume of the system down. The motor is a permanent magnet brushless motor with 2 poles, thus requiring switching devices operating at a minimum of 2,750 hertz (Hz). This is based on a design at Honeywell for motors in excess of 200,000 rpm (requiring a minimum switching frequency of 3,333 Hz). Although the details of the motor and motor controller are proprietary, the motor controller and motor are reported to have efficiencies of greater than 90 percent and greater than 93 percent, respectively, for a combined efficiency near 85 percent, which is truly remarkable at such speeds and frequencies. The motor stator and motor controller can be liquid- or air-cooled, but the rotor must be air-cooled. The cost does not include the entire cooling system cost, as the cooling system is shared with the vehicle's traction drive motor. Because of the critical importance of the CEM for HFCVs, the committee encourages continued support for further testing of this integrated subsystem—in particular with respect to noise and vibration as well as durability at conditions that can be expected in the automotive environment. In the opinion of the committee, this is the kind of stretch technology that is needed to reduce component size and material cost.

[21] B. James (Directed Technologies, Inc.), "Mass Production Cost Estimation of Direct H2 PEM Fuel Cell Systems for Automotive Applications," Presentation to the committee, October 26, 2009.

Recommendations

Recommendation 3-19. The Partnership should continue to focus on activities to reduce the cost, size, and losses in the power electronics and electrical machines.

Recommendation 3-20. The Partnership should conduct a project to evaluate the effect of battery charging on lithium-ion battery packs as a function of the cell chemistries, cell geometries, and configurations in the pack; battery string voltages; and numbers of parallel strings. A standardized method for these evaluations should be developed to ensure the safety of battery packs during vehicle operation as well as during plug-in charging.

Recommendation 3-21. The Partnership should consider conducting a project to investigate induction motors as replacements for the permanent magnet motors now almost universally used for electric propulsion.

STRUCTURAL MATERIALS

The challenge to the materials technical team is to generate a cost-neutral 50 percent vehicle weight reduction. The 50 percent weight reduction is critical to reaching FreedomCAR goals for fuel consumption and emissions. However, the target of no cost penalty for such a large weight reduction was unrealistic when set, and it remains unrealistic. A similar conclusion was stated in the Phase 2 report (NRC, 2008). What is missing at this juncture is a projection of what the cost penalty will likely be. For example, Berger et al. (2009) considered an aggressive weight-reduction program that yielded a 37 percent reduction (230 lb) in the Golf V body-in-white (BIW) which generated a 112 percent ($1,088) increase in cost. In other words, each 1 percent weight reduction in the BIW yielded a 3 percent increase in cost. Computer simulation was used to ensure that stiffness and crashworthiness requirements were met. An additional 50 percent weight savings of 115 lbs may be possible from downsizing brakes, suspension, engine, power train, and wheels and tires. But the associated cost savings are unlikely to make up the needed $1,000 plus. A full vehicle study is needed by FreedomCAR to assess the estimated overall cost penalty. Based on the above study, the outcome may be a penalty well over $500. What is also missing at this juncture is how the cost penalty changes as a function of the percent of weight reduction, assuming that the most effective mix of materials is used at each step in the weight-reduction process. This information will be needed in case the overall system-level targets for FreedomCAR need to be reset.

Weight Reduction Calculus

The impact of weight reduction on fuel consumption is well understood, and automotive OEMs have worked for many years to develop effective vehicle weight-reduction technologies. Consider, for example, a vehicle that is driven 12,000 miles per year having an average fuel economy of 25 mpg (0.04 gal/mi). The vehicle is then redesigned to achieve a 10 percent weight reduction using lightweight materials and/or better structural utilization. Fuel consumption would then be expected to be reduced by at least 6 percent, resulting in a new fuel economy of at least 26.6 mpg. The total of 480 gal used annually at 25 mpg would be reduced to 451 gal. For gas priced at $2.50/gal, the annual fuel costs of $1,270 at 25 mpg would be reduced by $72.50 due to the weight reduction. When computed over 6 years using an 8 percent discount rate for future savings, the resulting net present value (NPV) for the redesign is $335. Thus, there would be an NPV incentive of more than $100 to the buyer if the one-time, up-front added material costs were under $200.

This is an example of the value of developing weight-reducing technologies not only to the entire industry, but to the nation as well. For example, if each passenger car in the United States was reduced in weight by 10 percent, the expected annual savings in fuel would be more than 4.5 billion gallons, based on the 2006 vehicle fuel usage statistics. There would be large additional savings in societal value resulting from reduced pollution and reduced dependence on foreign oil that are not reflected in consumer (commercial) value of $289 computed above.

Also, there does not need to be a trade-off between increased fuel economy and safety. Mass reduction is an important means of improving fuel economy. But mass is not the same as size, and with efficient designs, low-mass cars can be made safe by improving crash-management design and reducing the frequency of accidents through improved accident-avoidance systems in vehicles and on highways.

Mass Decompounding

A weight-reduction process known as mass decompounding can be utilized when brakes, suspension, and power train can be redesigned to gain secondary weight savings as a result of the primary weight savings made in the structure through the use of lightweight materials. A lighter vehicle can perform equally well with smaller brakes, a less hefty suspension, and a smaller engine.

During the past year, the materials technical team arrived at a useful rule of thumb in which 1.0 to 1.5 lb of secondary weight savings should be achievable for each 1 lb of primary weight saved, provided that the entire vehicle can be redesigned to take advantage of the savings. The relationship for the amount of secondary weight savings was determined from an analysis of teardown data from two vehicle databases by the materials technical team.

If it is assumed that each 1.0 lb of lightweight material generates 1.25 lb of secondary weight savings on average, then a current vehicle weighing

3,000 lb will require 667 lb of primary weight savings to meet the 50 percent weight reduction goal of 1,500 lb.

Magnesium Power-Train Components

The targets for the project on magnesium power-train components have been to replace aluminum components with magnesium for a minimum weight savings of 15 percent and a cost penalty of less than $2.00 for each pound saved. This project was completed in September 2009. The mass reduction achieved was 29 percent for the magnesium components and 7.8 percent for the engine subsystem, which exceeded the weight-savings goal. The cost penalty was found to be $3/lb at current magnesium prices. Although over the cost target, the outcome was judged as demonstrating that magnesium was both technically feasible and potentially cost-effective in these applications.

Polyolefin Feedstock

Cost has been a limiting factor in the use of commercial carbon-fiber-reinforced composites in the design of automotive structures and body panels. Importantly, the materials technical team has shown that major cost savings appear possible through the use of polyolefin for the feedstock in making carbon fibers.[22] As with all lightweight materials applications, the trade-offs between cost and weight will need to be reevaluated as the price of oil changes. If the price of oil doubles, the cost of polyolefin will increase significantly, since polyolefin comes mostly from oil or perhaps natural gas. If the price of oil doubles, for example, the incentive to use carbon fibers in cars will increase because of the reduction in weight, but the cost of the polyolefin and therefore the carbon fiber will also increase. The DOE needs to understand the trade-offs there.

Recycling End-of-Life Vehicles

New materials present challenges to recycling. ANL is developing a one-fifth scale pilot operation to assess how to recover residual metals and polymers from the residue from shredders. A second pilot operation is to begin in the first quarter of 2010 to evaluate the recycling of polyurethane foams by converting them to polyols. The FreedomCAR and Fuel Partnership also needs to consider how to recycle carbon-fiber-reinforced composites including carbon-fiber hydrogen tanks.

---

[22] See <http://www1.eere.energy.gov/vehiclesandfuels/pdfs/merit_review_2009/light-weight_materials/lm_05_warren.pdf>.

Response to the Phase 2 Recommendations

A recommendation from the Phase 2 report (NRC, 2008, p. 9) is as follows:

> Recommendation. The materials research funding should largely be redistributed to areas of higher potential payoff, such as high-energy batteries, fuel cells, hydrogen storage, and projects associated with infrastructure issues. However, materials research for projects that show a high potential for enabling near-term, low-cost mass reduction should continue to be funded.

The response of the Partnership was as follows:[23]

- Strategic lightweighting is an important enabler for reducing fuel consumption.
- We recommend that lightweighting materials research funding not be redistributed to support other technology areas.
- We agree that materials research for projects that show a high potential for enabling near-term, low-cost mass reduction should continue to be funded.

Recommendations

The materials needed to make the required weight reductions—high-strength steels, aluminum, titanium, magnesium, and fiber-reinforced composites—are available. The key issue is not improving their performance but getting the weight reductions needed at an acceptable cost. The structural materials efforts and budget should reflect this reality. The resources required in the future may be less or more if major pilot programs are needed. Systems analysis is an approach that can highlight where critical cost reductions are needed. The polyolefin feedstock is a good example of what can be achieved. The high cost of aerospace-quality carbon fiber is a major impediment to achieving cost-effective compressed hydrogen storage.

Recommendation 3-22. The materials technical team should develop a systems-analysis methodology to determine the currently most cost-effective way for achieving a 50 percent weight reduction for hybrid and fuel cell vehicles. The materials team needs to evaluate how the cost penalty changes as a function of the percent weight reduction, assuming that the most effective mix of materials is used at each step in the weight-reduction process. The analysis should be updated on a regular basis as the cost structures change as a result of process research breakthroughs and commercial developments.

[23] See J. Quinn (GM) and J. Carpenter (DOE), "Materials Tech Team Peer Review Report," Presentation to the committee, August 4, 2009, Southfield, Michigan.

Recommendation 3-23. The magnesium castings study is completed, and no further technical effort is anticipated by the Partnership as recommended in the Phase 2 report. However, magnesium castings should be considered in completing the cost reduction recommendation listed above.

Recommendation 3-24. Methods for the recycling of carbon-reinforced composites need to be developed.

REFERENCES

Anderson, B., and I. Wade, 2001. Metal resource onstraints for electric vehicle batteries, *Transportation Research Record* D 6: 297-324.

ANL (Argonne National Laboratory). 2009. *Technical Assessment of Cryo-Compressed Hydrogen Storage Tank Systems for Automotive Applications.* Report ANL/09-33, by R. K. Ahluwalia, T. Q. Hua, J.-K. Peng, Argonne National Laboratory, Argonne, Illinois; S. Lasher, K. McKenney, and J. Sinha, TIAX LLC, Cambridge, Massachusetts. Argonne, Ill.: ANL. Federal Grant Number DE-ACO2-06CH11357. Available on the Web at <http://www1.eere.energy.gov/hydrogenandfuelcells/pdfs/cyro_compressed_auto.pdf>.

Bandivadekar, A., K. Bodek, L. Cheah, C. Evans, T. Groode, J. Heywood, E. Kasseris, M. Kromer, and M. Weiss. 2008. *On the Road in 2035: Reducing Transportation's Petroleum Consumption and GHG Emissions.* Report No. LFEE 2008-05 RP (July). Cambridge, Massachusetts: MIT Laboratory for Energy and the Environment.

Berger, Lutz et al., 2009. "Super LIGHT CAR—the Multi-Material Car Body." Presented at the 7th European LS-DYNA Conference. Available on the Web at <http://www.dynalook.com/european-conf-2009/B-V-03.pdf>.

Department of Energy (DOE). 2009a. *Hydrogen, Fuel Cells and Infrastructure: Multi-Year Research, Development and Demonstration Plan.* DOE/GO-102003-1741. Washington, D.C.: U.S. Department of Energy, Energy Efficiency and Renewable Energy (original edition, 2004; revised April 2009). Available on the Web at <http://www.eere.energy.gov/hydrogenandfuelcells/mypp/>.

DOE. 2009b. Hydrogen Program. "Announcement for H Prize." Available on the Web at <http://www.hydrogen.energy.gov/news_h-prize_administrator.html>.

DOE. 2009c. *Actions and Evidence Report, April 2.* Submitted to the National Research Council (NRC) Committee on Review of the FreedomCAR and Fuel Research Program, Phase 3, documenting responses by DOE to recommendations in the Phase 2 report [see NRC, 2008].

DOE. 2009d. 2009 *Annual Progress Report, DOE Hydrogen Program.* DOE/GO-102009-2950 (November). Washington, D.C.: Office of Energy Efficiency and Renewable Energy. Available on the Web at <http://hydrogendoedev.nrel.gov/pdfs/progress09/2009_doe_hydrogen_program_annual_progress_report_cover.pdf>.

El-Rafaie, A., and F. Johnson. 2009. "Scalable, Low-Cost, High Performance IPM Motor for Hybrid Vehicles." Presentation at the DOE Annual Merit Review, May 22. Available on the Web at <http://www1.eere.energy.gov/vehiclesandfuels/pdfs/merit_review_2009/advanced_power_electronics/ape_08_elrefaie.pdf>.

Howell, D. 2009. *Annual Progress Report for 2008, Energy Storage Research and Development (January).* Washington, D.C.: U.S. Department of Energy, Office of Energy Efficiency and Renewable Energy. Available on the Web at <http://www1.eere.energy.gov/vehiclesandfuels/pdfs/program/2008_energy_storage.pdf>.

James, B., and J. Kalinoski. 2009. "Mass-Production Cost Estimation of Automotive Fuel Cell Systems." Presentation at DOE 2009 Annual Merit Review, May 21, Arlington, Virginia. Available on the Web at <http://www.hydrogen.energy.gov/pdfs/review09/fc_30_james.pdf>.

Kim, G., and K. Smith. 2008. "Three-Dimensional Lithium-Ion Battery Model." 4th International Symposium on Large Lithium Ion Battery Technology and Application, Tampa, Florida, May 12-14. NREL/PR-540-43166. Available on the Web at <http://www.nrel.gov/vehiclesandfuels/energystorage/pdfs/43166.pdf>.

Nozawa, N., T. Maekawa, S. Nozawa, and K. Asakura. 2009. *Development of Power Control Unit for Compact-Class Vehicle*, SAE 2009-01-1310, April.

NRC (National Research Council). 2008. *Review of the Research Program of the FreedomCAR and Fuel Partnership, Second Report*. Washington, D.C.: The National Academies Press.

Rogers, S. 2009. "Advanced Power Electronics and Electric Machines." Presentation at the DOE Annual Merit Review, May 21. Available on the Web at <http://www1.eere.energy.gov/vehiclesandfuels/pdfs/merit_review_2009/advanced_power_electronics/ape_0_rogers.pdf- 916.8KB>

Satyapal, S. 2009. "Hydrogen Program Overview." DOE Annual Merit Review and Peer Evaluation Meeting, May 18. Available on the Web at <http://www.hydrogen.energy.gov/pdfs/review09/program_overview_2009_amr.pdf>.

Sinha, J., S. Lasher, and Y. Yang. 2009. "Direct Hydrogen PEMFC Manufacturing Cost Estimation for Automotive Applications." Presentation at the DOE 2009 Annual Merit Review, May 21, Arlington, Virginia. Available on the Web at <http://www.hydrogen.energy.gov/pdfs/review09/fc_31_sinha.pdf>.

Taylor, J. 2009. "Development, Test and Demonstration of a Cost-Effective, Compact, Light-Weight, and Scalable High Temperature Inverter for HEVs, PHEVs, and FCVs." DOE Annual Merit Review and Peer Evaluation Meeting, May 21. Available on the Web at <http://www1.eere.energy.gov/vehiclesandfuels/pdfs/merit_review_2009/advanced_power_electronics/ape_07_taylor.pdf>.

TIAX. 2009. *Technical Assessment of Compressed Hydrogen Storage Tank Systems for Automotive Applications*. Report to United States Department of Energy, Office of Energy Efficiency and Renewable Energy; Hydrogen, Fuel Cells and Infrastructure Technologies Program; Part 1 (Volume 1), December 10. Federal Grant Number: DE-FC36-04GO14283, Stephen Lasher, Kurtis McKenney, and Jayanti Sinha, TIAX LLC, Cambridge, Massachusetts, Reference: D0268; Rajesh Ahluwalia, Thanh Hua, and J-K Peng, Argonne National Laboratory.

Weiss, M. A., J.B. Heywood, E.M. Drake, A. Schafer, and F.F. AuYeung. 2000. *On the Road in 2020: A Life-Cycle Analysis of New Automobile Technologies*. Cambridge, Massachusetts: MIT Laboratory for Energy and the Environment.

Xu, J., H.R. Thomas, Rob W. Francis, Ken R. Lum, Jingwei Wang, Bo Liang. 2008. A review of processes and technologies for the recycling of lithium-ion secondary batteries, *Journal of Power Sources* 177 (2008) 512-527.

ANNEX

TABLE 3A-1 Technical System Targets: Onboard Hydrogen Storage for Light-Duty Vehicles

Storage Parameter	Units	2010	2015	Ultimate
System Gravimetric Capacity				
Usable, specific-energy from H_2	kWh/kg	1.5	1.8	2.5
(net useful energy/max system mass)[a]	(kg H_2/kg system)	(0.045)	(0.055)	(0.075)
System Volumetric Capacity				
Usable energy density from H_2	kWh/L	0.9	1.3	2.3
(net useful energy / max system volume)	(kg H_2/L system)	(0.028)	(0.040)	(0.070)
Storage System Cost[b]	$/kWh net	4	2	TBD
(and fuel cost)[c]	($/kg H_2)	(133)	(67)	
	$/gge at pump	2-3	2-3	2-3
Durability/Operability				
Operating ambient temperature[d]	°C	–30/50 (sun)	–40/60 (sun)	–40/60 (sun)
Min/max delivery temperature	°C	–40/85	–40/85	–40/85
Cycle life (1/4 tank to full)[e]	Cycles	1000	1500	1500
Cycle life variation[f]	% of mean (min) at % confidence	90/90	99/90	99/90
Min delivery pressure from storage system; FC = fuel cell, ICE = internal combustion engine	Atm (abs)	4FC/35 ICE	3FC/35 ICE	3FC/35 ICE
Max delivery pressure from storage system[g]	Atm (abs)	100	100	100

continued

TABLE 3A-1 Continued

Storage Parameter	Units	2010	2015	Ultimate
Charging/Discharging Rates				
System fill time (for 5 kg H_2)	Min (kg H_2/min)	4.2 min (1.2 kg/min)	3.3 min (1.5 kg/min)	2.5 min (2.0 kg/min)
Minimum full flow rate	(g/s)/kW	0.02	0.02	0.02
Start time to full flow (–20 °C)[h]	s	5	5	5
Start time to full flow (20 °C)[h]	s	15	15	15
Transient response 10%-90% and 90%-0%[i]	s	0.75	0.75	0.75
Fuel Purity (H_2 from storage)[j]	% H_2	99.99 (dry basis)	99.99 (dry basis)	99.99 (dry basis)
Environmental Health and Safety				
Permeation and leakage[k]	Scc/h	Meets or exceeds applicable standards	Meets or exceeds applicable standards	Meets or exceeds applicable standards
Toxicity	—			
Safety	—			
Loss of usable H_2[l]	(g/h)/kg H_2 stored	0.1	0.05	0.05

[a] Generally the "full" mass (including hydrogen) is used, for systems that gain weight, the highest mass during discharge is used.

[b] 2003 US$; total cost includes any component replacement if needed over 15 years or 150,000 mile life. The storage system costs are currently under review and will be changed at a future date.

[c] 2005 US$; includes off-board costs such as liquefaction, compression, regeneration, etc; based on H_2 production cost of $2 to $3/gasoline gallon equivalent untaxed, independent of production pathway.

[d] Stated ambient temperature plus full solar load. No allowable performance degradation from –20C to 40C. Allowable degradation outside these limits is TBD.

[e] Equivalent to 100,000; 200,000; and 300,000 miles respectively (current gasoline tank spec).

[f] All targets must be achieved at end of life.

[g] For delivery the storage system, in the near term, the forecourt should be capable of delivering 10,000 psi (700 bar or ca. 70 MPa) compressed hydrogen, liquid hydrogen, or chilled hydrogen (35 to 77 K) and up to 5,000 psi (350 bar or ca. 35 MPa). In the long term, it is anticipated that delivery pressures will be reduced to between 50 and 150 atm for solid state storage systems, based on today's knowledge of sodium alanates.

[h] Flow must initiate within 25% of target time.

[i] At operating temperature.

[j] The storage system will not provide any purification, but will receive incoming hydrogen at the purity levels required for the fuel cell. For fuel cell systems, purity meets SAE J2719, Information Report on the Development of a Hydrogen Quality Guideline in Fuel Cell Vehicles. Examples include: total nonparticulates, 100 ppm; H_2O, 5 ppm; total hydrocarbons (C1 basis), 2 ppm; O_2, 5 ppm; He, N_2, Ar combined, 100 ppm; CO_2, 1 ppm; CO, 0.2 ppm; total S, 0.004 ppm; formaldehyde (HCHO), 0.01 ppm; formic acid (HCOOH), 0.2 ppm; NH_3, 0.1 ppm; total halogenates, 0.05 ppm; maximum particle size, <10 μm; particulate concentration, <1 μg/L H_2. These are subject to change. See Appendix C on Hydrogen Quality, to be updated as fuel purity analyses progress. Note that some storage technologies may produce contaminants for which effects are unknown; these will be addressed as more information becomes available.

[k] Total hydrogen lost into the environment as H_2; relates to hydrogen accumulation in enclosed spaces. Storage system must comply with CSA/NGV2 standards for vehicular tanks. This includes any coating or enclosure that incorporates the envelope of the storage system.

[l] Total hydrogen lost from the storage system, including leaked or vented hydrogen; relates to loss of range.

SOURCE: DOE (2009a).

4

Hydrogen and Biofuels

The FreedomCAR and Fuel Partnership was originally focused on power systems driven by hydrogen fuel cells but now is examining three power system approaches: fuel cells using hydrogen, advanced combustion engines using biofuels, and plug-in hybrid electric vehicles (PHEVs) and battery electric vehicles (BEVs) using electricity. This chapter reviews programs focused on providing the fuels needed by the first two of these approaches while minimizing petroleum imports and greenhouse gas emissions. The issue of interfacing between the nation's electricity transmission and distribution system to provide the electricity needed for PHEVs and BEVs is addressed in Chapter 2, "Crosscutting Issues."

Hydrogen is an energy carrier produced from a variety of energy sources as discussed in this chapter and is the fuel that makes vehicular fuel cells feasible. Biofuels, energy carriers for solar energy and thus renewable fuels, are produced from a variety of biological sources, such as plant materials or algae. Programs on each of these approaches are reviewed in this chapter.

HYDROGEN PRODUCTION, DELIVERY, AND DISPENSING

As discussed in Chapter 1, the FreedomCAR and Fuel Partnership in the Department of Energy's (DOE's) Office of Energy Efficiency and Renewable Energy (EERE) includes the hydrogen production, delivery, and dispensing program, which is in turn part of the Hydrogen, Fuel Cells, and Infrastructure Technologies program (HFCIT; now called the Fuel Cell Technologies [FCT] program). This program addresses a variety of means of producing hydrogen in distributed and centralized plants using technologies that can be made available in the short, medium, and long term. The manager of the FCT program is the overall DOE hydrogen program manager. There are three fuel technical teams: fuel pathway integration, hydrogen production, and hydrogen delivery, with participation from the DOE and five energy companies that joined the Partnership 5 years ago. The technical teams report to the Fuel Operations Group, consisting of energy directors and DOE program managers, who in turn report to the Executive Steering Group.

A number of important programs related to FCT are carried out in other parts of the DOE. Work on growing, harvesting, transporting, and storing biomass as well as work on using solar heat to produce hydrogen are also carried out in the EERE but are not part of the Partnership.[1] The Office of Fossil Energy (FE) supports the development of technologies to produce hydrogen from coal and related carbon-sequestration technologies. The Office of Nuclear Energy (NE) supports research on the potential use of nuclear heat to produce hydrogen, and the Office of Science (SC) supports fundamental work on new materials for hydrogen storage, catalysts, and fundamental biological or molecular processes for hydrogen production, as well as work potentially affecting other areas of the FreedomCAR and Fuel Partnership.

As discussed elsewhere in this report, the DOE recently added two utility partners to the Partnership to address issues associated with emergence of PHEVs and BEVs. With time this should bring additional attention to the issues associated with providing the required electricity while increasing energy security and reducing greenhouse gas (GHG) emissions.

In reviewing the hydrogen production, delivery, and dispensing area, the committee considered whether it is appropriate for the federal government to be involved, and without exception the committee concluded that government involvement is appropriate and needed. As will be shown in this chapter, the DOE through the FCT program continues to make substantial progress, ensuring that hydrogen can be made available to meet the needs of fuel-cell-powered vehicles as they emerge. Continued work is needed to minimize cost and GHG emissions and reduce dependence on natural gas. Although the current abundance and low cost of natural gas make it attractive as a transition source of hydrogen, reducing dependence on natural gas should remain a long-term objective for hydrogen production.

HYDROGEN FUEL PATHWAYS

The hydrogen fuel/vehicle pathway integration effort is charged with looking across the full hydrogen supply chain from well (source) to tank. Specifically, the goals of this integration effort are to (1) analyze issues associated with complete hydrogen production, distribution, and dispensing pathways; (2) provide input to

[1] This EERE program is coordinated with programs in the Department of Agriculture. See on the Web <http://www.usda.gov/wps/portal/!ut/p/_s.7_0_A/7_0_1OB?navid=ENERGY&navtype=MS>.

the Partnership on goals for individual components; (3) provide input to the Partnership on needs and gaps in the hydrogen analysis program including the important industrial perspective; and (4) foster full transparency in all analyses, including an independent assessment of information and analyses from other technical teams. This effort involves source-to-vehicle-tank analysis, including costs, energy use, safety, availability of critical resources, and carbon dioxide (CO_2) emissions.

The accomplishment of these goals is overseen by the fuel pathways integration technical team (FPITT), with representation from the DOE, the energy companies, and the National Renewable Energy Laboratory (NREL). FPITT's expertise supports the analysis efforts of the Partnership, coordinates fuel activities with the vehicle systems analysis technical team, recommends additional pathway analyses, provides input from industry on practical considerations, and acts as honest broker for the information generated by other technical teams.

The DOE continues to make important progress toward understanding and preparing for the transition to hydrogen fuel. In the continuing source-to-wheels analysis, seven pathways, including both distributed and centralized hydrogen production, have been assessed, and the key drivers for pathway costs, energy use, and emissions have been identified. In addition, estimates have been developed for the water, electricity, natural gas, and platinum requirements for various pathways, and a biomass supply-and-demand assessment for major U.S. cities and regions was developed. A hydrogen quality study by the Argonne National Laboratory (ANL) was reviewed, and efforts are underway to incorporate hydrogen quality, cost, and benefit into the pathway analysis protocol. This will be very important, given that different pathways produce hydrogen with different levels of impurities that significantly impact performance and perhaps durability. The Society of Automotive Engineers (SAE) and the International Organization for Standardization (ISO) have developed standards for hydrogen purity that should be finalized in 2010. These standards can be further modified if research indicates that different standards are justified.

The technology is available to produce and distribute hydrogen commercially for large users, but it is not yet completely optimized and cost-effective for supplying local vehicle fueling stations. Research efforts are focused on the further development of options that reduce cost, dependence on imported petroleum and natural gas, and greenhouse gas emissions. The primary constraint to the broad availability of hydrogen is the construction of a distribution system similar to the natural gas pipeline network. The Partnership has already developed several options for distributed hydrogen generation that could be used while such a national distribution system is being built.

As indicated above, the long-term effort of the Partnership has thus far been focused on hydrogen. However, the Partnership now is examining three power system approaches, only one of which involves hydrogen: fuel cells powered by hydrogen, advanced combustion engines powered by biofuels, and PHEVs and BEVs powered by electricity. Clearly, additional effort is needed to develop meaningful comparisons of the fuel implications of these three approaches. As the fuel pathways integration technical team has indicated, biomass resource information is inconsistent and often outdated. And further definition is needed of the effects of significant penetration of PHEVs and BEVs in the market on electricity generation and distribution, including the impacts on the use of imported oil or gas and emissions of greenhouse gases as well as criteria pollutants.

Recommendation 4-1. The DOE should broaden the role of the fuel pathways integration technical team (FPITT) to include an investigation of the pathways to provide energy for all three approaches currently included in the Partnership. This broader role could include not only the current technical subgroups for hydrogen, but also subgroups on biofuels utilization in advanced internal combustion engines and electricity generation requirements for PHEVs and BEVs, with appropriate industrial representation on each. The role of the parent FPITT would be to integrate the efforts of these subgroups and to provide an overall perspective of the issues associated with providing the required energy in a variety of scenarios that meet future personal transportation needs.

HYDROGEN PRODUCTION

The hydrogen production program embodies hydrogen generation from a wide range of energy sources as a means of enhancing U.S. energy security and reducing GHG emissions. The hydrogen production technical team facilitates the development of commercially viable technologies through nonproprietary dialogue among the commercial and federal sectors to guide program efforts. Energy sources under study include natural gas, coal, biological systems, nuclear heat, wind, solar heat, and grid-based electricity; grid-based electricity employs several types of energy sources to varying extents, depending on geographical area. Direct comparisons of the costs and other consequences of using these approaches are not included here because the technologies are at different stages of development and the adequacy of domestic reserves of the resources varies, as pointed out in the National Research Council's Phase 2 report (NRC, 2008). In addition, hydrogen purity varies depending on the production approach used. Since hydrogen quality can affect fuel cell performance and durability, the cost of removing impurities to provide equivalent hydrogen may vary and influence the comparisons to some extent.

The hydrogen production program includes both long-term and short-term approaches. In the short term, when a hydrogen pipeline system is not in place, distributed generation in relatively small plants will be required to supplement hydrogen available from existing, large-scale commercial plants. As the fleet of fuel-cell-powered cars grows and hydrogen demand increases, centralized hydrogen-generation plants with pipeline distribution will become increasingly attractive and are expected to partially replace distributed generation.

The rest of this section reviews the DOE's ongoing programs on hydrogen production involving thermal, electrolytic, and photolytic processes (DOE, 2010).

Thermal Processes

Approaches to hydrogen generation using thermal processes include coal and biomass gasification, reforming of bio-derived fuels, and high-temperature thermochemical splitting of water. The DOE's program to improve natural gas reforming, completed in 2009, has established the feasibility of distributed generation at fueling stations using reforming and has directionally improved gas cleanup technologies for centralized plants. Commercial options now exist to generate hydrogen either in distributed or centralized plants using natural gas.

Hydrogen Production from Coal and Biomass

This subsection addresses the application of two domestic feedstocks, coal and biomass, to the manufacture of hydrogen. The production of hydrogen from coal or from biomass feedstocks appears in the Hydrogen Production Roadmap (DOE, 2009a,b) as both a midterm technology (coal gasification with carbon sequestration) and a long-term technology (biomass gasification with carbon sequestration). The most critical challenges to the use of either feedstock are (1) the capital cost of the gasification processes and (2) the cost and availability of carbon sequestration.

Both feedstocks, coal and biomass, share a generally similar gasification process[2] that has been in commercial use for nearly a century—the solid feedstock is gasified by reacting it with just enough oxygen to increase its temperature so that steam can react with the remaining carbonaceous material to produce "syngas," a mixture of carbon monoxide (CO) and H_2. The syngas is then cleaned to remove contaminants—such as particles, sulfur, ammonia, and mercury—and further processed to improve the ratio of H_2 to CO by using the water–gas shift reaction (NAS/NAE/NRC, 2009). A wide slate of products can be produced, but those of chief interest here are hydrogen and CO_2. In addition, electric energy could be sold as a by-product, possibly offsetting some of the cost of hydrogen production.

Advantages and Limitations of Hydrogen Production from Coal and Biomass. Both feedstocks, coal and biomass, offer abundant, domestic resources for the manufacture of hydrogen. In the case of coal, most estimates suggest a resource sufficient to meet the needs of the United States for the next century at current

[2] The distinct chemical and physical properties of each feedstock require special adaptation and so present challenges when coal and biomass are co-fired in a single gasification process.

rates of consumption. However, use of the coal resource for hydrogen production will be paced by the availability of carbon sequestration—currently the secure disposal of the CO_2 in underground reservoirs and possibly in the future disposal as a solid through advanced technologies.

In contrast, biomass, though renewable, is limited by the sustainability of the land, water, and chemical resources required for its production. Further, the attractiveness of the biomass feedstock would be increased markedly if the CO_2 that it emits could also be permanently sequestered from the biosphere.

Thus CO_2 disposal becomes an essential goal in the use of both coal and biomass for hydrogen. Yet for all its importance, a full-scale demonstration of permanent geologic storage has not been made within the U.S. legal, regulatory, and social framework, even though it has been demonstrated in a few locations around the world.

Funding for carbon-sequestration research has grown steadily from about $69 million in FY 2006 to about $150 million in FY 2009. The American Recovery and Reinvestment Act of 2009 also allotted about $3 billion for projects to demonstrate this technology.[3] The first U.S. demonstration, FutureGen, was discontinued in June 2008 but was started again with a new set of industrial partners a year later. The Department of Energy signed an agreement with its new industrial partners in the FutureGen Alliance to fund a reevaluation of the project for a 2010 go/no-go decision. Through regional partnerships in the United States, the DOE is planning to conduct 20 different tests by 2020. Worldwide, the target is 68 projects by 2020. This announced schedule implies that confirmation of the acceptability of the geological sequestering of CO_2 could not be complete before 2020.

In addition to geologic and terrestrial sequestration research and development (R&D) activities, the DOE also supports research on novel and advanced concepts that pursue chemical and biological methods of consuming CO_2. Examples of chemical methods include capturing CO_2 by reaction with magnesium sulfate to form carbonate, or formation of CO_2 clathrate; examples of biological methods include microbial conversion of CO_2 to methane or other hydrocarbons.

In summary, the commercial deployment of coal-to-hydrogen production prior to the availability of publicly acceptable CO_2 disposal will have adverse effects on CO_2 emissions. The committee believes that at best the demonstration of CO_2 sequestration is unlikely to see completion before 2020, and the record of similar projects suggests that it might well be later.

Furthermore, the availability of biological feedstocks from sustainable sources is essential for biomass gasification to become a major producer of hydrogen.

Recommendation 4-2. The DOE's Fuel Cell Technologies program and the Office of Fossil Energy should continue to emphasize the importance of dem-

[3] L. Miller, DOE, "Status and Outlook for Carbon Capture and Storage," Presentation to the committee, October 26, 2009, Washington, D.C.

onstrated CO_2 disposal in enabling essential pathways for hydrogen production, especially for coal.

Recommendation 4-3. The Fuel Cell Technologies program should adjust its Technology Roadmap to account for the possibility that CO_2 sequestration will not enable a midterm readiness for commercial hydrogen production from coal. It should also consider the consequences to the program of apparent large increases in U.S. natural gas reserves.

Recommendation 4-4. The EERE should continue to work closely with the Office of Fossil Energy to vigorously pursue advanced chemical and biological concepts for carbon disposal as a hedge against the inability of geological storage to deliver a publicly acceptable and cost-effective solution in a timely manner. The committee also notes that some of the technologies now being investigated might offer benefits in the small-scale capture and sequestration of carbon from distributed sources.[4]

Recommendation 4-5. The DOE should continue to evaluate the availability of biological feedstocks for hydrogen in light of the many other claims on this resource—liquid fuels, chemical feedstocks, electricity, food, and others.

Reforming of Bio-Derived Fuels

Before the demand for hydrogen is large enough to support large centralized production facilities, smaller distributed hydrogen generation at fueling station sites is expected to be the preferred option. The steam reforming of natural gas and water electrolysis are both technically attractive options for this. Neither process, however, provides a clear and practical renewable pathway. It is not practical to capture the CO_2 from the distributed reforming of natural gas, and the electricity generated for use in electrolysis is dependent on the grid makeup, which on average releases large amounts of CO_2. The distributed reforming of bio-derived liquids such as ethanol, sugars, or bio-oils can provide a renewable option for distributed hydrogen generation in the early stages of a hydrogen fuel buildup. A recent study concluded that, from a technical standpoint, up to 2 million barrels per day of gasoline-equivalent fuel could be produced from biomass available in 2020 but that the actual level of production could be achieved some time beyond that, in about 2035 (NAS/NAE/NRC, 2009).

A wide range of bio-derived liquids can be reformed into hydrogen, but ethanol, being the largest produced biofuel, has received the most development attention. The process for steam reforming ethanol into hydrogen is similar to

[4] A description of these advanced concepts appears on the Web at <http://fossil.energy.gov/programs/sequestration/novelconcepts/index.html>.

that for steam reforming natural gas except for higher temperatures. In this sense steam reforming of ethanol requires little further development, but it still has some unique challenges that can be addressed. The primary challenges involve catalyst activity and coking and the overall cost of the hydrogen produced. Since the processes for the steam reforming of natural gas and ethanol are similar, it is expected that any process-related technology advances in one can be applied to the other. However, the feedstock cost issues are very different.

The overall process of first producing ethanol from a cellulosic source, then transporting the ethanol to a station, and then reforming this into hydrogen results in a hydrogen cost that is higher than that for reforming natural gas. Its applicability then will be related to the relative costs of ethanol compared with that of natural gas and also to the value associated with the renewable aspect of hydrogen production. A tax on CO_2 emissions would favor hydrogen from any biomass feedstock, including cellulosic ethanol reforming. It is possible, or even likely, that future state or federal regulations will encourage or mandate that a percentage of hydrogen be made in a renewable fashion. California already has such a program.[5]

Whereas distributed natural gas reforming has demonstrated the ability to meet the hydrogen cost targets of $2.00 to $3.00 per gallon gasoline equivalent (gge) (based on the DOE standard set of assumptions), distributed ethanol reforming has not. The current cost estimates are higher than $4.00/gge, and the targets for 2014 and 2019 are $3.80/gge and $3.00/gge, respectively.

To meet these targets, further improvements are needed in catalyst performance, process design cost aspects, and feedstock cost reductions. All of these issues are being investigated, with indications of progress. The DOE is investigating using other bioliquids such as sorbitol, glucose, glycerol, methanol, propylene glycol, and less refined sugars such as cellulose and hemicelluloses that may have potential for cost improvements over ethanol.

One promising technology path is aqueous-phase reforming that can process water-soluble carbohydrates such as glucose, sorbitol, glycerol, or methanol. The process conditions for aqueous-phase reforming are less severe than for the vapor-phase reforming that is used for natural gas or ethanol reforming. As a result, catalyst coking is not a significant problem as it can be for vapor-phase reforming. This technology is at a very early development stage and holds some promise for reducing costs. Laboratory batch experiments have indicated very high reactor conversion of cellulosic biomass (95 percent) at high hydrogen selectivity (74 percent).

In summary, there is likely to be a need for a renewable distributed-hydrogen-generation method. Reforming a bioliquid is a viable approach. There are several different feedstock and technology pathways to do this—for example, cellulosic ethanol with vapor-phase reforming and glucose with aqueous-phase reforming,

[5] See, for example, <http://www.energy.ca.gov/low_carbon_fuel_standard/>.

among many others. It will be very difficult to meet the hydrogen cost targets without a significant reduction in current costs for bioliquid.

Recommendation 4-6. The Partnership should prioritize the many biomass-to-biofuel-to-hydrogen process pathways in order to bring further focus to development in this very broad area.[6]

High-Temperature Thermochemical Splitting of Water

The DOE is funding six projects that address the high-temperature technique of thermochemical water splitting for the centralized production of hydrogen. To split water directly by brute force requires temperatures of about 2000°C. By using various chemical cycles, the reaction temperatures can be reduced to the 500°C to 1100°C range. These temperatures can be achieved by many means, including next-generation high-temperature nuclear reactors or solar concentrators. Most solar design concepts for this centralized production method use power towers to get the high powers and high temperatures required. Some of the concepts also use electricity to power a still-elevated but much lower-temperature electrolysis step.

Most of the work done so far in the area of thermochemical water splitting has been funded by the Office of Nuclear Energy and has been on the sulfur-iodine cycle. It is not clear to the committee that all attractive chemical cycles have been identified. Several projects are nearing completion, and this provides an opportunity to review and down-select projects to identify promising approaches.

The committee understands that the Office of Nuclear Energy will not be funding chemical cycles for hydrogen production in the future. In the view of the committee, the EERE's effort to identify solar, thermochemical approaches for future funding would be enhanced by carrying out a systems analysis of candidate systems after conducting a workshop to ensure that all promising, potential cycles have been identified. It would also be useful to see if any attractive options are evolving in the DOE Office of Basic Energy Sciences (BES) program, including the new Energy Frontier Research Centers.

Recommendation 4-7. The Partnership should consider conducting a workshop to ensure that all potentially attractive high-temperature thermochemical cycles have been identified, and it should carry out a systems analysis of candidate systems to identify the most promising approaches, which can then be funded as money becomes available.

Recommendation 4-8. The EERE funding for high-temperature thermochemical cycle projects has varied widely and was very low in FY 2009. The committee

[6] N. Gupta, "Hydrogen Production Technical Team," Presentation to the committee, Slide 11, Biomass Processes, August 5, 2009, Southfield, Michigan.

believes that these centralized production techniques are important, and thus adequate and stable funding for them should be considered.

Electrolytic Processes

This subsection covers programs aimed at splitting water using electricity in an electrolysis process. The coupling of wind power with electrolysis is included as one embodiment of this technology.

In the long term, water electrolysis represents a significant option for hydrogen generation in the development of a refueling infrastructure. Its attractiveness stems from the fact that (1) it is relatively simple compared to alternative methods; (2) it can positively impact carbon emissions if powered by renewables; (3) it can generate relatively pure hydrogen, potentially at elevated pressure, thereby making downstream cleanup processing simpler and reducing compression requirements; and (4) its efficiency is largely independent of unit size. Furthermore, water electrolysis can be placed at the "point of use," allowing it to satisfy regional hydrogen supply needs, while at the same time it has the potential for large-scale centralized operations. In the near term the process is attractive, as it can facilitate a proof-of-concept fueling option because water electrolysis technology and systems are available today. However, water electrolysis is still a small segment of the total hydrogen-generation capacity, because capital costs are high and have not been seriously reduced to date, and operating costs are high; together these costs lead to high-cost hydrogen. Except for selected military and industrial-based uses, electrolysis systems have been built without volume and cost-reduction benefits due to the lack of high-volume manufacturing. Even with these limitations, conventional water electrolysis exhibits high efficiencies (percentages in the 70s versus lower heating value [LHV]) and long lifetimes using multiple chemistries and processes, and electrolysis systems are available commercially in low- and high-volume production of gas.

The primary R&D activities are focused on component or engineering (balance of plant) enhancements. Such advancements have the potential to impact the energy requirements that are over and above what is needed to split the water from the perspective of a fundamental electrochemistry requirement. Examples of R&D activities include new membranes for both acid and alkaline chemistries, hardware and configuration changes, and advanced catalysis. Specifically, membrane development could possibly impact the resistance of the stack, thereby reducing the ohmic losses if the membrane has enhanced conductivities over those of currently used materials. In all cases, such energy-consumption-reduction and capital-cost-reduction efforts will have to succeed without seriously impacting the efficiency and lifetimes currently exhibited in commercial and military electrolyzers. Lastly, such developments must make progress against the goals for hydrogen cost. Presented in Table 4-1 are the DOE cost targets for distributed hydrogen generation from water electrolysis.

TABLE 4-1 DOE Cost Status and Targets for Distributed Hydrogen Generation from Water Electrolysis, 2006, 2012, 2017

Characteristics/Units	2006 Status	2012 Target	2017 Target
Hydrogen cost, \$/kg H_2	4.80	3.70	<3.00
Electrolyzer capital cost, \$/kg H_2	1.20	0.70	00.30
Electrolyzer efficiency, %			
Based on LHV	62	69	74
Based on HHV	73	82	87

NOTE: LHV, lower heating value; HHV, higher heating value.
SOURCE: See DOE's *Multi-Year Research Development and Demonstration Plan: Planned Activities for 2005-2015* (updated April 2009). Available on the Web at <http://www1.eere.energy.gov/hydrogenandfuelcells/mypp/pdfs/production.pdf>.

Higher-risk, longer-term new technologies have been recently reported (Farmer, 2009) and include both high-temperature electrolysis and the photoelectrolysis of water. High-temperature electrolysis has advantages in the reduction of energy requirements to split water owing to the lower voltage requirements to dissociate water, but it requires high-temperature materials that are challenging. The high-temperature option, using solid oxide technology, also has the potential to make use of waste thermal energy, thereby making nuclear power plants attractive locations for centralized generation.

Another aspect of the long-term potential of the electrolysis process is that there are variations of the engineering configurations, making it even more attractive by possibly reducing system complexity. For example, in selected cases, hydrogen may be generated at substantial pressures, thereby reducing the need for follow-on mechanical compressors (DOE, 2005). Furthermore, from a final cleanup perspective, as electrolysis generates relatively pure hydrogen, the final cleanup stage from a pressurized system may use alternative, existing know-how to remove residual oxygen and moisture efficiently (e.g., high-pressure electrolysis followed by passive membrane separation). In such cases, the electrolyzer may be able simply to generate pressure and purity without significantly impacting electrical consumption and capital. If such is accomplished, refueling hydrogen may be available in remote and non-methane-accessible regions that meet hydrogen purity specifications.

The DOE recognizes that water electrolysis may play an important role in the hydrogen infrastructure, and the DOE is supporting numerous electrolysis efforts related to capital, electrocatalytic processes, and configuration and engineering. In addition, a number of systems analyses now include water electrolysis.

Furthermore, photoelectrolysis has the potential to improve the efficiency of water splitting, but fundamental research is needed to establish the feasibility of this approach relative to conventional electrolysis (DOE, 2008).

Recommendation 4-9. Water electrolysis should remain an integral part of the future hydrogen infrastructure development. The DOE should continue to fund novel water electrolysis materials and methods, including alternative membranes, alternative catalysts, high-temperature and -pressure operation, advanced engineering concepts, and systems analysis. Additional efforts should be placed on advanced integration concepts in which the electrolyzer is co-engineered with subsequent upstream and downstream unit operations to improve the overall efficiency of a stand-alone system.

Recommendation 4-10. Commercial demonstrations should be encouraged for new designs based on established electrolytic processes. For newer concepts such as high-temperature solid oxide systems, efforts should remain focused on laboratory evaluations of the potential for lifetime and durability, as well as on laboratory performance assessments.

Wind- and Solar-Driven Electrolysis

The DOE continues to study at NREL opportunities to couple wind and solar energy with electrolysis, and it has several projects to improve the efficiency of electrolyzers. The program has recently demonstrated about 70 to 71 percent efficiency at the stack level. Higher-pressure electrolyzers could be a thrust for the future and could reduce the compression energy for storage and vehicle refueling.

The hydrogen storage can be used to offset at least in part the intermittent and variable nature of the wind and solar resource. This approach can be employed with three different energy pathways: wind to grid; wind to electrolysis unit to hydrogen; and hydrogen to fuel cell to grid. These outputs can be varied if there is not enough demand for hydrogen for vehicle fueling.

Some of the challenges with a wind- and solar-driven electrolysis approach include efficient power electronics for dc-to-dc and ac-to-dc conversion, and controllers and communications protocols to match the source to the electrolyzer. Additional valuable data and experience can be gained by continuing the operation and upgrades of the facility at NREL.

Recommendation 4-11. Work on close coupling of wind and solar energy with electrolysis should be continued with stable funding. Further improvements in electrolyzers, including higher stack pressure, and in power electronics will benefit this application.

Photolytic Processes

This subsection covers the discussion of programs oriented toward using solar energy to split water. Included are biological and photoelectrochemical hydrogen production.

The production of hydrogen using microorganisms, utilizing energy by absorbing incident light and nutrients, can be a carbon-neutral process. The FreedomCAR and Fuel Partnership identifies four main biological production pathways: photolytic (direct water splitting), photosynthetic bacterial (solar-aided organic decomposition), dark fermentative (organic decomposition), and microbial-aided electrolysis (electric power-aided organic decomposition) (DOE, 2009b). The commercial viability of these processes is highly uncertain, and the Partnership classified this approach as "long term." The activity has been supported by the BES. There are many barriers to technical success that, if overcome, would result in a process competitive with other pathways for hydrogen production. Thus, a possible application identified for this approach is to generate hydrogen from dilute feedstock in waste streams from other processes that would not be captured otherwise. The technical barriers include, among others, lack of information on microorganisms with suitable characteristics for biological hydrogen production; efficiency in light utilization; efficiency in feedstock utilization; cost; and product purity. In spite of the difficulties, the Partnership reported some noticeable progress—for example, the successful cloning of the Tla2 gene to enable 15 percent absorbed solar-to-chemical-energy conversion efficiency in microalgae. Internationally, this approach is pursued actively.

Photoelectrochemical water splitting, utilizing electrolysis, converts solar energy directly into chemical energy in the form of hydrogen.[7] A semiconductor material is used to collect light energy and produce hydrogen and oxygen using electrolysis. This also is supported by the BES and is classified as "long term." Barriers to technical success are found in the semiconductor materials used to capture light energy, the photochemical device, the integration of the device into an operating system, and the development of the storage needed to compensate for the diurnal light cycle.[8] Given these barriers, this could be the highest-risk approach currently in the program.

Barring spectacular breakthroughs, the potential impact of biological and photoelectrochemical hydrogen production will be limited and far in the future. Support of this approach has been by BES, which is appropriate because of its exploratory nature and because discoveries could just as likely have applications other than for the Partnership.

The committee finds no clearly defined targets or vision of the photolytic approach that will contribute to the overall hydrogen production goals, and as a result it is unclear whether the Partnership should retain this approach in its portfolio of activities.

---

[7] Although interesting work is being done with the use of microorganisms and photoelectrochemical techniques to generate hydrogen, and R&D is expected to continue, it is too early to consider them as viable options in the context of Partnership goals.

[8] Alternatively, hydrogen could be stored for use when there is no sunlight. This would also be a barrier, given the issues with hydrogen storage discussed in this report.

Recommendation 4-12. The Partnership should examine the goals for the photolytic approach to producing hydrogen using microorganisms and formulate a vision with defined targets. Otherwise, this approach should be deemphasized as an active research area for hydrogen production.

HYDROGEN DELIVERY, DISPENSING, AND TRANSITION SUPPLY

A significant factor in fuel cost and (source-to-wheels) efficiency for fuel-cell-powered vehicles is the means for delivering, storing, and dispensing hydrogen, especially compared to the petroleum delivery system, which is low-cost and efficient. In a fully developed hydrogen economy, the postproduction part of the supply system for high-pressure hydrogen will probably cost as much as production and consume as much energy (NRC/NAE, 2004). The distribution costs are of even greater concern in the transition period when there is a lack of demand, particularly when hydrogen from centralized production is available. In such cases distribution could easily cost more than production.

Dispensing systems for gaseous hydrogen must be designed to prevent excessive temperature increases in the vehicle tank during pressuring and filling, particularly for 700 bar (approximately 70 MPa or 10,000 psi) operation. As a result, communication between the vehicle and the refueling dispenser is required so that pressure and temperature can be monitored and controlled.

As pointed out in the National Research Council's Phase 2 report (NRC, 2008), there are five main ways to deliver hydrogen from centralized production to refueling stations: pipeline, liquid, gas containers, one-way liquid carriers, and two-way liquid carriers. Given the importance of the area, all of these have been studied in the program.

The DOE program on the delivery, storage, and dispensing of hydrogen is comprehensive and includes aggressive cost targets (see Table 4-2). The goal is to reduce the delivery and dispensing cost to less than $1 per kilogram of hydrogen by 2017. This compares to current costs of $3-$5/kg at low volume and $2-$3/kg at high volume. Given that all of the physical steps involved in delivery and dispensing have been practiced for decades by the gas industry, the committee continues to question whether it will be possible to reduce costs to the target levels, but clearly significant cost reductions are very important to the outlook for hydrogen-powered vehicles.

Funding of this important program has been variable. Funding of $1.1 million in FY 2006 and $6.3 million in FY 2007 was followed by $9.5 million in FY 2008 and $3.3 million in FY 2009. In spite of this inconsistent funding, progress has continued to be made. Cost has been reduced from $3-$5/kg to roughly $2-$3/kg through advances in pipelines, tube-trailers, and liquefaction.[9]

---

[9] M. Gardiner and J. Kegerreis, "Hydrogen Delivery Technical Team," Presentation to the committee, August 5, 2009, Southfield, Michigan.

TABLE 4-2 Cost Targets for Hydrogen Delivery and Dispensing ($ per kilogram of hydrogen)

Activity	2010	2012	2015	2017
Delivery from central plant to refueling gate		<0.90		<0.60
Dispensing at refueling site[a]	<0.80		<0.40	

[a] Includes compression/storage; centralized H_2 available at 300 psi.
SOURCE: NRC (2008), p. 98.

Progress has been made in all areas of the program. Delivery models have been developed that predict delivery and dispensing costs for different methods as a function of market penetration. Hydrogen compression has been directionally advanced by investigating a centrifugal compressor design and also electrochemical compression. Promising aging studies on a fiber-reinforced polymer pipe material were completed. In addition, studies of carbon-fiber composites and glass fibers indicate that the capacity of tube trailers can be increased by a factor of two to three, leading to a potential cost reduction for these trailers, according to DOE estimates, of up to 50 percent. The plan for FY 2010 includes $4.5 million for this area.

Past funding in the area of hydrogen delivery and dispensing has not been based on program needs but apparently on budget constraints. Reducing the cost of delivery and dispensing from the current $2 to $3 per kilogram of hydrogen to the 2017 target of less than $1 per kilogram of hydrogen will require substantial and consistent funding based on program needs. Otherwise, any chance of meeting the 2017 target will be forgone.

Recommendation 4-13. Hydrogen delivery, storage, and dispensing should be based on the program needed to achieve the cost goal for 2017. If it is not feasible to achieve that cost goal, emphasis should be placed on those areas that would most directly impact the 2015 decision regarding commercialization.[10] In the view of the committee, pipeline, liquefaction, and compression programs are likely to have the greatest impact in the 2015 time frame.[11] The cost target should be revised to be consistent with the program that is carried out.

[10] The program framework is based on a 2015 target date for getting all of the technical information needed by the automotive manufacturers to make decisions on commercialization of fuel-cell-powered vehicles. Presumably, these decisions will involve in part the assurance of the availability of hydrogen at a sufficient number of locations to provide the fuel needed.

[11] Compression will take on even greater importance as the delivery pressure is raised from 350 to 700 bar (about 5,000 to 10,000 psi, or 35 to 70 MPa).

BIOFUELS FOR INTERNAL COMBUSTION ENGINES

Liquid hydrocarbon fuels made from biomass and used in an internal combustion engine (ICE) are another pathway that can have an effect on oil imports and CO_2 emissions. This pathway differs from the other two primary pathways being developed by the Partnership in that the light-duty vehicle (LDV) and drive system do not require new technology for this pathway to grow. The hydrogen fuel cell pathway and the battery electrification pathway (hybrid electric vehicle [HEV], PHEV, and BEV) both require new drivetrain technology for them to begin and to grow. Biomass-derived ethanol and biodiesel are already available in the marketplace for use with today's ICE cars. The technology development needs are almost all related to producing the biofuel rather than to the use of the biofuel in an ICE.

About 8 billion gallons per year (BGY) of biofuels were produced in the United States in 2008, with most of this being ethanol made from corn. This amount is about 5 percent of the total gasoline use measured by volume, or a little less than 4 percent measured by energy content. Less than 10 percent of this 8 BGY is biodiesel made primarily from soy and waste oils. The total is planned to increase to 36 BGY by 2022, as outlined in the Renewable Fuel Standard that is part of the Energy Independence and Security (EISA) Act of 2007 (Public Law 110-140). Most of the increase beyond the present is scheduled to be from non-grain-based sources. Corn-based ethanol could grow to 12 BGY by 2012 but is not anticipated to grow beyond this.

To accommodate these plans, a number of challenging barriers must be resolved with the biomass production, logistics, conversion into biofuel, distribution of the biofuel, and end use of the biofuel:

1. *Biomass production*—The rapid growth of biofuels has come from the use of corn and soy for producing ethanol and biodiesel. To meet the future EISA targets, movement to second-generation sources (crop and forest residues) and even third-generation sources (energy crops such as perennial grasses, fast-growing trees, and algae) will be necessary.
2. *Feedstock logistics*—Harvesting, storing, preprocessing, and delivering the biomass to a conversion facility can cost as much as 20 percent of the total cellulosic ethanol cost. Innovative business models and new technologies are needed to reduce these costs.
3. *Conversion to biofuel*—Currently, cellulosic ethanol production and other biofuel technologies needed to reach the EISA targets are too expensive to compete in the marketplace. Cellulosic ethanol conversion technology is in the large-pilot-plant or small-demonstration-plant phase of development.
4. *Biofuel distribution*—Much of the land area needed for increasing biofuel production will be in the Midwest, and by inference much of the biofuel production also, although much of the demand for biofuels will

be on the coasts. A much larger system of trucks, trains, barges, blending and storage terminals, and perhaps pipelines and station storage will be needed.

5. *Biofuel end use*—As ethanol use increases to meet EISA targets, the average gasoline blend will need to contain more than the current 10 percent maximum of ethanol (E10). Although some of today's LDVs can use E85, the ability to move the entire LDV fleet to E20 and above is needed prior to achieving the 2022 targets.

Biofuels and the FreedomCAR and Fuel Partnership

Within the DOE, the Biomass Program[12] has the responsibility for managing the development and progress for the bulk of the needs for biofuels, including the needs for biomass production, feedstock logistics, and biomass conversion to a biofuel. Historically the DOE focused on biofuel distribution and end use through the FreedomCAR and Fuel Partnership (DOE, 2010). This split of focus puts responsibility for making biofuels with the Biomass Program and the responsibility for delivering the biofuel and the LDV drivetrain responsibility with the FreedomCAR and Fuel Partnership. As this committee is reviewing just the Partnership and not the entire biofuel program, its comments apply to those areas for which FreedomCAR has responsibility.

The split of focus described above appears logical and takes advantage of the capabilities within the Partnership developed over the last several years for systems analysis and the ongoing ICE development work. Just as the systems analysis of the hydrogen infrastructure has better defined the challenges, barriers, and possible solutions to implementing a hydrogen infrastructure, so too can a thorough analysis of the biofuel infrastructure identify barriers, possible solutions, and costs for distributing biofuels.

A key assumption with regard to the use of biofuels is that all near- and mid-term biofuels must be fungible with existing liquid fuels and existing distribution infrastructure (DOE, 2010). Blending biofuels with gasoline and diesel as is done with ethanol and biodiesel has distinct advantages that hold down distribution costs by using a portion of the existing distribution system and takes advantage of the existing car technology and large gasoline and diesel markets for sales. Doing this creates impacts on the existing petroleum fuel distribution system and on vehicle drivetrain performance.

E85 can be viewed as an example of how the biofuel affects the existing petro-

---

[12] Algal biofuels are now receiving increased attention and are included in the DOE Aquatic Species Program of the Biomass Program (http://www1.eere.energy.gov/biomass/pdfs/algalbiofuels.pdf). With further research this could become a viable long-term option for producing biodiesel, gasoline, or other fuels. One other possible long-term option is the use of photosynthetic bacteria to produce hydrocarbons from CO_2.

leum system and the engine technology. Because of the high octane of ethanol (129 Research Octane Number [RON] and 102 Motor Octane Number [MON] and 116 [R+M]/2), a flexible fuel engine is designed to perform with both a regular 87 (R+M)/2 octane gasoline and the much-higher approximately 105 (R+M)/2 octane E85 so that it could use either fuel. If, however, the engine were optimized only for the high-octane E85, it could be made more efficient. From a petroleum refinery perspective, E85 is a fuel with excess octane giveaway, because the fuel has much higher octane than the gasoline specification. High-octane components of gasoline are much more expensive than are lower-octane components. If the 15 percent of petroleum gasoline components to the blend were optimized to take advantage of the excess octane, a lower-cost E85 could result.

This example illustrates some of the complexities of integrating three large industries into an overall efficient system. The choice of the biofuel impacts both the design of the ICE and the makeup of the petroleum fraction blended with the biofuel. In the short term, ethanol is the biofuel of choice, as it is the only biofuel in the market now in a significant amount. Much of the research on biomass conversion, however, is focused on the next generation of biofuels, such as mixed alcohols, biobutanol, green gasoline, and diesel. A close collaboration between the three industries in the Biomass Program and the Partnership in analyzing the overall system should help ensure the best overall system design.

As the emphasis for biofuel growth beyond that anticipated for ethanol is likely to come from a new biofuel or one that is blended with gasoline or diesel fuel, it creates an opportunity to improve the ICE overall efficiency through a fuel and engine optimization program. Furthermore, there is a need for a thorough systems analysis of the biofuel distribution and end-use system that accounts for engine technologies and petroleum-blending fuel properties could help to identify priority areas for further development. Such development could result in modified priorities for different biomass sources, conversion processes, biofuels, distribution systems, and engines.

Recommendation 4-14. A thorough systems analysis of the complete biofuel distribution and end-use system should be done. This should include (1) an analysis of the fuel- and engine-efficiency gains possible through ICE technology development with likely particular biofuels or mixtures of biofuels and conventional petroleum fuels, and (2) a thorough analysis of the biofuel distribution system needed to deliver these possible fuels or mixtures to the end-use application.

REFERENCES

DOE (U.S. Department of Energy). 2005. Small Business Innovation Research and Small Business Technology Transfer Programs, Phase II Grant Abstracts, FY 2005, DOE Grant DE-FG02-04ER83905. Washington, D.C.

DOE. 2008. *Annual Progress Report, DOE Hydrogen Program.* Section II-B, Electrolysis. Washington, D.C.: Office of Energy Efficiency and Renewable Energy.

DOE. 2009a. *Hydrogen Production, Overview of Technology Options (January).* FreedomCAR and Fuel Partnership. Available on the Web at <http://www1.eere.energy.gov/hydrogenandfuelcells/pdfs/h2_tech_roadmap.pdf>.

DOE. 2009b. *Hydrogen Production Roadmap, Technology Pathways to the Future* (January). FreedomCAR and Fuel Partnership. Available on the Web at <http://www1.eere.energy.gov/hydrogenandfuelcells/pdfs/h2_production_roadmap.pdf>.

DOE. 2010. *Biomass Multi-year Program Plan (April).* Office of Biomass Energy, Office of Energy Efficiency and Renewable Energy. Available on the Web at <http://www1.eere.energy.gov/biomass/pdfs/mypp.pdf>.

Farmer, R. 2009. "Hydrogen Production." DOE Annual Merit Review and Peer Evaluation Meeting, May 19. Arlington, Virginia. Available on the Web at <http://www.hydrogen.energy.gov/pdfs/review09/pd_0_farmer.pdf>.

NAS/NAE/NRC (National Academy of Sciences/National Academy of Engineering/National Research Council). 2009. *Liquid Transportation Fuels from Coal and Biomass: Technological Status, Costs, and Environmental Impacts*. Washington, D.C.: The National Academies Press.

NRC (National Research Council). 2008. *Review of the Research Program of the FreedomCAR and Fuel Partnership, Second Report.* Washington, D.C.: The National Academies Press.

NRC/NAE (National Research Council/National Academy of Engineering). 2004. *The Hydrogen Economy: Opportunities, Costs, Barriers, and R&D Needs*. Washington, D.C.: The National Academies Press.

5

Overall Assessment

This chapter presents an overall view of progress in the FreedomCAR and Fuel Partnership since the previous, Phase 2 review by the National Research Council (NRC, 2008). It delineates many of the major achievements made in each of the technical areas under investigation and also identifies the critical technical barriers that still need to be overcome. It also addresses the overall adequacy and balance of the Partnership by a review of and comments on the level of budgetary resources and effort being directed at each of the major budget line items. Finally, the committee's concluding comments are presented.

MAJOR ACHIEVEMENTS AND TECHNICAL BARRIERS

Although there is evidence of steady progress in every technical area being pursued by the Partnership, there remain very formidable barriers to the mass-production of affordable hydrogen and fuel-cell-powered vehicles. Further, it is also becoming more obvious that major barriers can involve more than one technology. An example of this is the difficulty in arriving simultaneously at the emergence of the affordable, consumer-friendly fuel cell vehicles and the emergence of a widespread network of hydrogen refueling stations offering affordable hydrogen fuel. Even though the need for, as well as the difficulty in achieving, this simultaneous emergence has been well known from the beginning of FreedomCAR, most efforts were appropriately focused on advancing specific technologies. Clearly, without resolving the enabling technologies for the ultimate production of consumer-acceptable fuel cell vehicles and the production and distribution of affordable hydrogen, the simultaneous emergence issue would have been moot. Rigorous scenario modeling might serve to illuminate the simultaneous emergence problem by allowing the qualitative testing of alternate futures. The quantitative implications of these futures might be understood through simple and transparent spreadsheet models.

Advanced Combustion and Emissions Control

Higher efficiency and reduced emissions as compared to those associated with current internal combustion engine (ICE) vehicles will be very important to the success of biofuels, advanced hybrid electric vehicles (HEVs), and even plug-in hybrid electric vehicles (PHEVs). In addition, the use of hydrogen fuel in ICEs can offer an alternative for expanding the availability of hydrogen for refueling vehicles prior to the widespread distribution of fuel cell vehicles. There is also the potential for homogeneous charge compression ignition (HCCI) engines to provide even better combinations of efficiencies and emissions than either diesel or spark-ignited ICEs. Such advances can be implemented by a better understanding of fundamental processes, and such is the primary orientation of these efforts. However, almost all aspects of engine operation are being pursued by the Partnership.

Among the accomplishments in this area were the following:

- The demonstration of a peak brake thermal efficiency of 43 percent for conventionally fueled and 45 percent for hydrogen-fueled ICEs has taken place. Engines with these efficiencies operating in hybrid vehicles could result in system efficiencies approaching those of fuel cell vehicles.
- The development of a predictive model for spark-assisted HCCI combustion is complete. This could help lead to the elusive solutions for successfully controlling HCCI engines.

The fundamental research being performed by the advanced combustion and emission control (ACEC) technical team is generating the knowledge base necessary to identify how to optimize the combustion process at any operating condition. This understanding is being incorporated into detailed computational fluid dynamic (CFD) simulations, which in turn accurately replicate the experimental results with minimum adjustable numerical tuning (Ge et al., 2010).

Primary barriers include the following:

- In all of these endeavors, the key barrier continues to be the need for detailed fundamental understanding of the chemical, thermal, and physical processes taking place within the power train and combustion system.
- Also, as with almost all technologies being pursued, cost is a barrier. Specifically, the technical team has had difficulty specifically addressing its cost target. The team has assumed that the base engine cost will be

$20/kW and that the incremental cost for the technology improvements, which includes enhanced aftertreatment, will be $10/kW. The team has not been able to confirm these estimates with public domain data.

Electrochemical Energy Storage

Energy storage is essential for any type of vehicle that recovers part of the kinetic energy that would otherwise be lost to heat dissipation during braking (regeneration) and/or operates without the primary energy converter (e.g., ICE or fuel cell) for some or all of its operation; these vehicles would include HEVs, PHEVs, battery electric vehicles (BEVs), and hydrogen-fueled fuel cell vehicles (HFCVs). While there are many ways to store energy (flywheels, compressed air, elastomers, hydraulic springs, batteries, capacitors, etc.), the Partnership has focused on electrochemical storage—primarily batteries and to a lesser extent ultracapacitors. Batteries can serve as primary energy sources onboard the vehicles as well as being a means of recovering kinetic energy, unlike some of the other energy storage technologies that serve better as short-term power devices.

The emphasis on batteries is even greater with the Obama administration, which has announced a goal to "put 1 million plug-in hybrid cars—cars that can get up to 150 miles per gallon—on the road by 2015 . . ." (Satyapal and Davis, 2009). Conceptually, PHEVs are very similar to HEVs but require far more battery energy, enough to provide an all-electric-design driving range of (typically) 10 to 40 miles. At this time, versions of lithium-ion (Li-ion) batteries with different chemistries seem to be the most likely candidates, and these dominate most Partnership battery research activities.

Among the accomplishments are the following:

- Three Li-ion battery chemistries classified by the cathode material, including (1) lithium nickel, cobalt and aluminum; (2) lithium iron phosphate; and (3) lithium manganese spinel and a carbon anode, have been developed and tested for HEV applications. Since the Phase 2 review, there has been improvement in discharge and regenerative pulse power rating, calendar life as measured by accelerated testing, and increased cycle life. This applies to all three of the chemistries discussed. Laboratory data indicate that energy and power density requirements for HEVs will be met. With only Daimler having a production HEV with Li-ion batteries, the real cost is still unknown. However, recent announcements about the Nissan Leaf indicate that the cost and durability for a BEV, which is a much tougher application, may be within reach.
- Safety is an important factor in the design of the battery. To determine safety, the Partnership has conducted extensive abuse testing, including mechanical (crushing, perforation, and shock), electrical (external shorting, overcharging, and overdischarging), and thermal (over-temperature

from external and internal sources). The three Li-ion chemistries were tested at both the cell and battery-pack level in an attempt to access their readiness for use and to improve their design and manufacturability.

Barriers include the following:

- The system battery cost for an HEV is still very high, about twice the target of \$500.
- For the PHEV battery research, the Partnership has followed multiple paths of developments, also using different materials and designs to optimize performance, life, and cost. At present none of the battery chemistries meets the performance, life, or cost goals for 2012 requirements.

Electrical Propulsion and Electrical Systems

BEVs, HEVs, PHEVs, and fuel-cell-powered vehicles have in common a dramatic increase in electrical and electronic components as compared to those in conventional ICE vehicles. All of these vehicles are completely or partially electrically driven. Therefore, the performance and costs of components such as electric drive motors and power electronics are major factors for a range of advanced vehicles.

The Partnership has appropriately focused on key technical areas that are precompetitive, with the objective of long-term reductions in size (volume and weight) and cost. To accomplish this, emphasis has been on better packaging, cooling, materials, and devices.

Among the accomplishments for power electronics are the following:

- Good progress is being made on a scalable inverter. The concept for a scalable inverter is that it can easily be scaled to meet different power levels. This is a cost-shared contract provided by DOE that runs from October 2007 through March 2011, with Delphi as the project lead contractor.
- The Partnership is investigating a new process for making silicon carbide (SiC) on silicon substrates. SiC devices are better than those based on doped silicon since they operate at higher temperatures and have faster switching times. Potentially they lead to smaller and more efficient power electronics.
- Successful experiments for a novel approach using direct backside cooling for the thermal management of power electronics have been conducted. More effective thermal management can lead to reduced weight and volume as well as an increase in life.

Some key accomplishments toward better permanent magnet electric drive motors are as follows:

- The design of a 30 kW continuous (55 kW peak) motor is making good progress. Key objectives are for scalable motors to meet the very tough performance specifications and the use of novel soft magnetic material with a tripling of resistivity enhancement, and some specific accomplishments within this project are these:
  —*Motor design.* Two rotor and two stator concepts have been developed and analyzed in detail.
  —*Low-loss soft magnetic materials.* Bulk amorphous alloy compositions were identified as well as a kilogram-scale production by gas atomization. Also, a novel microstructure developed to enhance resistivity and magnetic properties, including composite soft magnetic material with doubling of resistivity enhancement, was demonstrated. Some cost and performance issues remain with these materials. Thus, while the materials are not satisfactory replacements for current materials, simply developing such low-loss materials is an important accomplishment.
  —*Low-loss permanent magnet materials.* Using a hydrogen-based route for processing the high-energy-density magnet materials used in electrical machines, the project has demonstrated a novel composite microstructure to minimize eddy current losses and seems very promising.
- Other accomplishments include the apparently successful testing of a new compressor expander motor (CEM) for fuel cell vehicles. The CEM project incorporates a high-speed drive (165,000 rpm) with a fairly low projected cost. The motor is a permanent magnet brushless motor based on a design at Honeywell for motors in excess of 200,000 rpm. Although details of the motor and motor controller are proprietary, the motor controller and motor are reported to have efficiencies of greater than 90 percent and greater than 93 percent, respectively, for a combined efficiency near 85 percent, which is truly remarkable at such speeds and frequencies.

Among the remaining significant barriers are the following:

- Permanent magnet motors still require rare-earth materials that are available from only a few countries and could become scarce and expensive or could become political or national security issues.
- Induction motors, while much less expensive and requiring no exotic materials, do not yet achieve efficiency levels sufficiently high for most vehicle (e.g., BEVs or PHEVs) drive motor applications.
- While much progress is being made on power electronics, silicon devices are still exclusively used. Perhaps in the future, cost-effective SiC devices capable of operating at higher temperature and faster will be developed, but there are not yet affordable options for higher-temperature and/or faster electronics.

Fuel Cells

Fuel cells are considered to have the potential for being the most efficient means of converting hydrogen fuel to useful power for vehicles. As such, they have been from the beginning of the program until very recently a primary focus of the FreedomCAR and Fuel Partnership. Progress has been steady, with an increase in almost every performance metric and a decrease in projected costs between each review. Even so, this technology is not yet fully developed, and additional advancements are needed. A noteworthy comment on significant achievements since the last review is that while almost every major target has been met in one form or another, they have so far been in separate initiatives and not from a collective single source.

Among notable achievements are these:

- Demonstrated stack lifetimes in on-road vehicles have increased from approximately 1,250 hours to 1,977 hours. With the goal of 5,000 hours, this represents a significant achievement since the Phase 2 NRC (2008) review.
- Single-cell and short-stack tests at the laboratory scale have demonstrated (using accelerated test protocols) much longer run times (3M Company, 7,200 hours) that, if demonstrated in vehicles under realistic on-road conditions, would meet or exceed the goals of the Partnership.
- Two separate DOE-funded studies, with independent oversight by industry experts, have concluded that at volumes of 500,000 units per year, the cost per kilowatt for the fuel cell subsystem including the fuel cell and balance of plant (BoP) will be approximately \$60-\$70/kW. These figures are still more than two times higher than the target but significantly lower than the \$107/kW presented during the Phase 2 review.
- The development of a membrane with double the conductivity and reduced in-plane swelling as compared to Nafion membranes (DOE, 2008).
- The development of improved non-invasive methods for visualizing water distribution inside fuel cells.[1]

Technical barriers that still remain include the following:

- For the fuel cell stack the technical barriers are the membrane and electrode life and cost. Neither cost nor durability targets have been met with a single technology.
- Previously, most cost-reduction efforts have been directed to the stack. However, the projected stack cost is now down to approximately the

[1] See DOE Annual Merit Review Meeting, May, 2009. Available on the Web at <http://www.hydrogen.energy.gov/annual_review09_proceedings.html>.

projected cost of the BoP. Consequently, to reach cost targets there must be appreciable reductions in BoP costs, which will probably require significant simplifications to the system.

Onboard Hydrogen Storage

Onboard hydrogen storage is a key enabler for fuel-cell-powered vehicles. A specific goal of the program is a vehicle driving range of greater than 300 miles between refuelings while simultaneously meeting vehicle packaging, cost, and performance requirements as well as those related to life-cycle issues, energy efficiencies, and safety with consideration of the possible environmental impact of implementing various hydrogen storage technologies.

The availability (or lack thereof) of an effective, affordable technology for onboard hydrogen storage was, and continues to be, a likely major issue in the ultimate decisions to mass-produce fuel cell vehicles. At the time of the Phase 2 report (NRC, 2008), only compressed hydrogen (and a few vehicles using liquid hydrogen) onboard storage systems had been utilized for Partnership experimental vehicles. The same is still true as this report is written.

Shortly before the Phase 1 report (NRC, 2005) was written, the Partnership had formed three new centers of excellence (COEs) for hydrogen storage. In the Phase 2 report (NRC, 2008), it was noted that the Chemical Hydrogen Storage COE had made significant progress in identifying materials with increased hydrogen storage capacity. The COEs have now investigated more than 350 candidate storage materials. Of those investigated, more than two-thirds (68 percent) have been discontinued, with about one-third still under investigation.

An additional center of excellence, the DOE Hydrogen Storage Engineering COE, was also established in the interim since Phase 2, to focus on systems issues that complemented the materials-oriented issues addressed by the other COEs. Another significant change by the Partnership since the Phase 2 report was to change (lower) the 2015 targets for system gravimetric density (hydrogen [H_2] storage weight/system weight from 9 percent to 5.5 percent) and volumetric density (kilowatt-hours of H_2 per liter of storage system volume, from 2.7 to 1.3). The system fill-time target was increased from 2.5 minutes for 5.0 kg H_2 to 3.3 minutes for 5.0 kg H_2.

Key achievements made since the Phase 2 NRC review include the following:

- A no-go decision made for vehicle hydrogen storage during the Phase 2 review was to discontinue applied research and development (R&D) in pure, undoped, single-walled carbon nanotubes based on the fact that they were not able to meet storage target of 6 wt% close to room temperature (2006).

- A no-go decision was made for sodium borohydride onboard vehicular hydrogen storage (2007).
- A down-select decision was made on chemical hydrogen storage materials (2008). Of 120 materials and classes of materials examined to date, 15 percent were selected for continued study.
- Metal hydride materials were down-selected. Of 74 materials investigated to date, 40 have been selected for further work.
- The Hydrogen Sorption COE has investigated 160 materials, and 35 percent are still in the COE's inventory. A down-select report is in preparation.
- Overall progress in system capacity is reported to have increased 50 percent since 2007.

Barriers remain formidable for onboard hydrogen storage. Specifically:

- The only complete system that has been successfully demonstrated in vehicles (other than a few liquid hydrogen demonstrations) is compressed hydrogen gas, which seems unlikely to meet any of the Partnership cost or performance targets.
- System weight and volume are too high for meeting the 300-mile vehicle driving range across a wide spectrum of vehicle platforms. Basically no suitable storage material has been identified and developed.
- The system cost is too high and needs to be able to compete with petroleum-based fuels. Cost areas include materials of construction and manufacturing methods and balance-of-plant components.
- Charging time (refueling) for material storage must meet consumer expectations. Storage capacity and rates of sorption and release need to be better understood and improved (the 2015 target is 3.3 minutes for 5 kg H_2 charge).
- The charging or discharging of hydrogen to a storage tank or material can be an energy consumer (requiring heating or cooling), which impacts the overall system efficiency.
- Systems issues need to be considered in addressing thermal management, durability and operability, hydrogen quality, containment vessels, dispensing technologies, and system life-cycle assessment and prediction.
- Codes and standards are needed for the entire system and for all interfaces.
- A better understanding of hydrogen sorption (physisorption and chemisorption) and desorption processes is needed.
- The cost of high-quality carbon fibers for high-pressure storage tanks needs to be reduced.

Materials

Reducing the mass of a vehicle will simultaneously reduce the required power for a given performance and increase fuel efficiency. Thus fuel cells and/or batteries (for fuel cell vehicles) and/or ICEs as well as batteries (for HEVs or PHEVs), power requirements, weight, volume, and cost of power plants can all be reduced with reduced vehicle mass.

Although there is undoubtedly some potential for mass reduction through design optimization using conventional materials, the automotive manufacturers have had mass reductions as priority goals for decades. This suggests that additional significant mass reductions are likely to be achieved through materials and corresponding fabrication changes. Among materials with the most promise are high-strength steel, aluminum, magnesium, and composites. With these, as well as other alternative materials, the biggest barrier is making components affordable as well as lighter.

Among materials, accomplishments of note are these:

- A project conducted for the possible substitution of magnesium for some aluminum components. This project was completed in September 2009. The mass reduction achieved was 29 percent for the magnesium components and 7.8 percent for the engine subsystem, which exceeded the weight savings goal. Although over the cost target, the outcome was judged as demonstrating that magnesium was both technically feasible and potentially cost-effective in some applications.
- The development of textile-based carbon-fiber precursors that could help lower the production costs of carbon fibers.
- The design and development of a vehicle magnesium front end.
- The development of better engineering property tools for tailored polymer composite structures.
- The development of a structural composites underbody.

The major barrier to using alternative materials to reduce vehicle mass is cost:

- The technology is available to replace many vehicle components with lighter alternatives. However, all alternative materials to date use base materials and/or production processes that are significantly more costly than are current materials and processes.

Fuel Infrastructure Technologies

The production, distribution, and delivery of affordable hydrogen is obviously essential if hydrogen is to become a widespread fuel available for refueling millions of fuel cells or other hydrogen-fueled vehicles. Even though relatively large quantities of hydrogen are produced in the United States (and the rest of the world), very little is distributed and used as vehicle fuel.[2] Consequently, there is virtually no effective infrastructure in place for large quantities of vehicle fuel. Thus, the Partnership R&D projects include essentially all phases necessary to develop a complete hydrogen vehicle fuel infrastructure.

Some of the more notable of the achievements are as follows:

- An analysis of infrastructure materials availability was completed, identifying key materials and their respective availability for an infrastructure for up to 10 million vehicles by 2025.
- A resource-availability analysis was completed that considered especially water and electrical energy.
- A refinement of a source-to-wheels analysis for better projecting both energy and emissions associated with various complete hydrogen pathways was made.
- A study on the feasibility of using glass fiber composite tanks to deliver cold hydrogen gas to refueling stations was completed.
- National and regional workshops involving hydrogen fuel station developers along with code officials were convened. The result was a Web-based information compendium to meet the needs and recommendations of the developers and authorities having jurisdiction. This is important because there are hundreds of federal, state, and local codes that developers must comply with to build widespread hydrogen refueling stations.

The primary barriers to hydrogen production, delivery, and refueling stations are given below:

- Costs for long-term hydrogen production with low or zero CO_2 emissions are a major issue. In the near term, natural gas can be used as a feedstock for producing relatively low cost hydrogen with relatively low CO_2 emissions, but the long-term availability of natural gas is not assured. For the long term, low-cost hydrogen with low to zero CO_2 production using non-fossil-fuel feedstocks is necessary, supplemented by coal if carbon capture and storage (CCS) technology is successfully developed. Long-term use of electrolysis depends on a substantial reduction of CO_2 emissions from the generation of grid electricity, which also is not assured.

[2] About 10 million metric tons of H_2 per year are produced in the United States, almost all of which is used for various industrial processes and, in fact, much of which never leaves the plants where it was produced. This amount (if it was available for fuel) could provide fuel for about 50 million fuel cell vehicles (DOE, 2005; NRC/NAE, 2004).

- Delivery has many issues to be resolved better, including capital costs for pipelines, compression, and on-site storage; efficiency for compression and for liquefaction; and the maintaining of hydrogen quality acceptable for fuel cells.
- There is a need for comprehensive codes and standards.

ADEQUACY AND BALANCE OF THE PARTNERSHIP

In two previous reports (NRC, 2005, 2008), the committee reviewed the funding for the FreedomCAR and Fuel Partnership and the allocation of that funding, both between hydrogen-related and non-hydrogen-related activities and between technologies perceived to be nearer-term and longer-term technologies. Generally speaking, the committee concluded in those earlier reports that the balance between technologies and between near and long term was appropriate. Major shifts in emphasis and funding have occurred over the past 12 months, and those shifts are explored in this section.

Since the beginning of the FreedomCAR program, and even earlier during the Partnership for a New Generation of Vehicles (PNGV) program, the NRC reviews have recommended government support emphasizing long-term, high-risk, high-payoff technologies. It was, and is, the view of the committee that this is an appropriate expenditure of government resources. However, current economic conditions, including the need for government support to prevent the collapse of two major automobile manufacturers, influence what the committee and the government consider "appropriate." It is still believed by the committee that support for long-term technologies such as the enablers for hydrogen to become a viable transportation fuel and the fuel cell R&D leading to affordable hydrogen fuel-cell-powered vehicles is very important and should be continued. Nonetheless, the committee agrees with government support for possible nearer-term technologies, especially those that could transfer some of the required transportation energy from petroleum to biofuels or to the electric power grid.

Historically, hydrogen-related activities represented approximately 70 percent of Partnership research and funding. This emphasis was consistent with the recommendations of the NRC report *The Hydrogen Economy: Opportunities, Costs, Barriers, and R&D Needs* (NRC/NAE, 2004) and the U.S. Department of Energy (DOE) report *Hydrogen Posture Plan: An Integrated Research, Development and Demonstration Plan* (DOE, 2004). It was also consistent with continuation of President Bush's commitment of $1.7 billion over 5 years (FY 2004 to FY 2008, the first 5 years of the Partnership). However, early in 2010, coincident with a new administration in Washington, D.C., all funding requests for hydrogen-related activities for vehicles were withdrawn. The stated reasons for this were that in order to achieve commercialization of hydrogen fuel cell vehicles, four major breakthroughs were required: namely, the sustainable production of hydrogen, effective distribution, onboard storage, and robust, reliable, low-cost fuel cells. It was apparently believed by the administration that the simultaneous achievement of these challenging tasks was highly unlikely in the 10- to15-year time frame, and consequently resources were redeployed to nearer-term activities.

Although the committee agrees that these four challenges to the deployment of HFCVs are indeed huge, it believes that the other two possible pathways to achieving the ultimate Partnership goals—namely, vehicles using biofueled ICEs and highly electrified vehicles (PHEVs and BEVs)—also face major challenges, and research on all three pathways deserves continued stable funding for the immediate future (see Appendix B, the committee's Interim Letter Report). Congress reached the same conclusion and reinstated most funding for hydrogen-related Partnership activities for FY 2010. The resulting FY 2010 funding for such activities is shown in Table 5-1. Figure 5-1 provides an estimate of the distribution of funding for the Hydrogen Program for FY 2009.

The companion Partnership activities under the general heading of the Office of Vehicle Technologies (VT) program continued to be adequately funded and in fact, the VT program, including similar activities in the separate 21st Century Truck Partnership, grew from $242 million in FY 2009 to $311 million in FY 2010, a 29 percent increase. This is illustrated in Table 5-2. (Note that the FY 2009 total is overstated by $31 million due to a one-time re-binning of three items normally included in hydrogen funding.) Figure 5-2 provides an estimate of the distribution of funding from the VT program in FY 2009.

The end result of these major swings in funding is that the hydrogen and fuel cell portion of the program currently represents some 50 percent of total funding, compared with roughly 70 percent in earlier years and 60 percent in FY 2009. Although the committee believes that a higher proportion of the funding could very justifiably be devoted to hydrogen and fuel cell activities, the nearer-term projects to which the majority of funding is now allocated are nevertheless well worthwhile, and much of that activity, such as improved batteries, more efficient electrical components, and lighter-weight materials, would also potentially benefit fuel cell vehicles in the future.

The DOE budget request for FY 2011 was submitted to Congress in February 2010. The committee strongly urges the DOE to maintain hydrogen-related funding at no less than the current level. Furthermore, this hydrogen activity should emphasize the most urgent needs to support a 2015 commercialization decision—namely, onboard storage, vehicle fuel cells, and distributed fueling. Not only is continued emphasis required to enable a robust 2015 decision, the fundamental objective on which the Partnership was founded, it is also precisely this type of high-risk/high-payoff research that is most appropriate for the use of public funds to augment private-sector spending.

Although not strictly comparable since the budget categories are somewhat different, the FY 2011 budget request is delineated in Table 5-3. Of course, it will be up to Congress as to where the final budget levels for FY 2011 end up. Also, Table 5-3 only provides estimates by the DOE of how the funding may

TABLE 5-1 Fuel Cell Technology and Related DOE Hydrogen Funding, FY 2009 and FY 2010 *(in thousands of dollars)*

Office/Activity	FY 2009	FY 2010
Hydrogen Production and Delivery R&D	10,000	15,000
Hydrogen Storage R&D	59,200	32,000
Fuel Cell Stack Component R&D	62,700	62,700
Technology Validation[a]	*See footnote a*	13,097
Transportation Systems R&D	6,600	3,201
Distributed Energy Systems R&D	10,000	11,410
Fuel Processor R&D	3,000	171
Safety, Codes and Standards[a]	*See footnote a*	8,839
Education[a]	*See footnote a*	2,000
Systems Analysis	7,713	5,556
Manufacturing R&D	5,000	5,000
Market Transformation	4,747	15,026
Total EERE	*168,960*	*174,000*
Fossil Energy	26,400[b]	~25,000[b,c]
Nuclear Energy	7,500	0
Science	38,284	~38,284[d]
DOE Total	241,144[a]	~237,284

[a] Under the Vehicle Technologies budget in FY 2009; the FY 2009 total noted for Fuel Cell Technology funding and related DOE Hydrogen funding does not include the three EERE Fuel Cell Technology activities moved to the Vehicle Technologies program in FY 2009. The DOE total including those activities is $272,633,000.

[b] Does not include funding for program direction.

[c] Includes coal to hydrogen and other fuels. Fossil Energy also plans $50 million for SECA in FY10.

[d] Exact funding for hydrogen- and fuel-cell-related projects to be determined. The Office of Science also plans approximately $14 million for hydrogen production research in the Office of Biological and Environmental Research in FY 2010. Projects in the Office of Basic Energy Sciences can be found on the Web at <http://www.hydrogen.energy.gov/pdfs/review09/2_kung_2009_amr.pdf>.

SOURCE: Submission to the committee by DOE, December 7, 2009.

be distributed, as noted in the footnotes to the table. Nevertheless, it does show continuing efforts on, among other activities, fuel cells for vehicles.

While these changes in hydrogen funding were taking place during 2009, another major initiative influencing Partnership goals emerged. That was the American Recovery and Reinvestment Act (ARRA) of 2009 (Public Law 111-5) and its massive funding of advanced technologies under the umbrella of economic stimulus. This expenditure is entirely separate from FreedomCAR and Fuel Partnership funding, but such initiatives as $1.5 billion for lithium-ion battery manufacture, $500 million for electric-drive component manufacturing, and $400 million for transportation electrification are clearly relevant to advanced development activities within the Partnership. The large ARRA expenditures

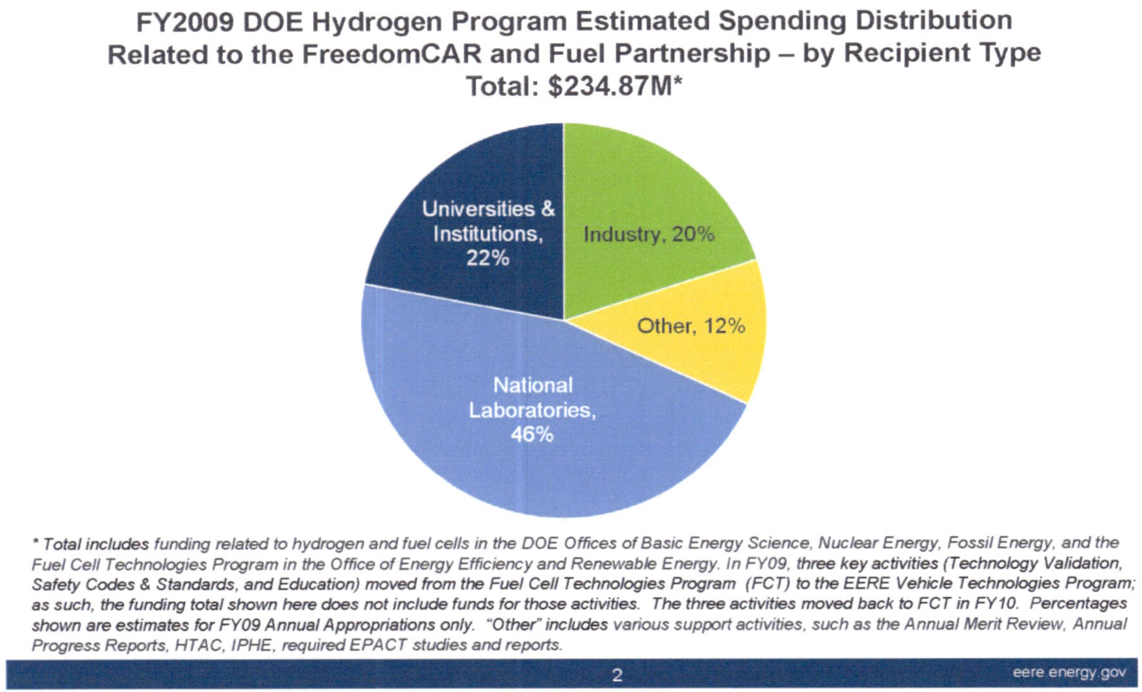

FIGURE 5-1 Distribution of funding from the Department of Energy Hydrogen Program for FY 2009.
SOURCE: Provided to the committee by DOE, April 23, 2010.

TABLE 5-2 Vehicle Technologies Program Funding, FY 2009 Appropriation and FY 2010 Estimate (in thousands of dollars)

Activity	FY 2009 FreedomCAR Appropriation	FY 2009 Total	FY 2010 FreedomCAR Estimate	FY 2010 Total
Hybrid Electric Systems	125,709	125,709[a]	140,960	145,733
Advanced Combustion Engines	25,427	40,800	33,990	57,600
Materials Technologies	28,256	39,903	38,355	50,723
Fuels Technologies	13,195	20,122	11,534	24,095
Technology Integration	2,700	46,704[a]	3,500	33,214
Total	195,287	273,238	228,339	311,365

[a] In FY 2009, the DOE transferred three activities from the Fuel Cell Technologies program to the Vehicle Technologies program. Technology Validation (FY 2009: $14,789) was included in Hybrid Electric Systems activity; Education (FY 2009: $4,200) and Safety, Codes, and Standards (FY 2009: $12,500) were included in Technology Integration. All three activities moved back to the Fuel Cell Technologies program in FY 2010.
SOURCE: Submission to the committee by the DOE, December 7, 2009.

could, at least temporarily, affect the appropriate levels of funding within the Partnership budget. The DOE should examine this and consider reallocations if warranted. Furthermore, late in 2009 (October 26), the DOE released a list of the first 37 approved research projects, totaling $151 million, under the Advanced Research Projects Agency-Energy (ARPA-E), a creation of the America Competes Act of 2007 (Public Law 110-69). On December 7, 2009, Energy Secretary Steven Chu announced a second round of funding amounting to $100 million for such projects. The total funds available for these ARPA-E projects are $400 million, drawn from the ARRA.[3]

Although many of the recipients of the ARPA-E funding are also participants in the FreedomCAR and Fuel Partnership, and staff from the two DOE program offices responsible for the Partnership assisted in the review of relevant ARPA-E proposals, these activities are not directly within the purview of Partnership leadership or this committee. However, they represent a substantial commitment of public funds to R&D and to the commercialization of technologies of major interest to the Partnership. Without passing judgment on the relative merits of all these ARRA and ARPA-E projects, it must be noted that, taken together, they overwhelmingly favor the "enhanced electrification of vehicles" pathway rather than either of the other two potential avenues, noted above, to the Partnership's (and the nation's) transportation energy goals. Indeed, a total of about $2.4 billion has been invested in battery manufacture, power electronics, and transportation electrification as well as $400 million for ARPA-E programs. These efforts are

[3] Details of these initial awards may be found on the Web at <http://arpa-e.energy.gov>.

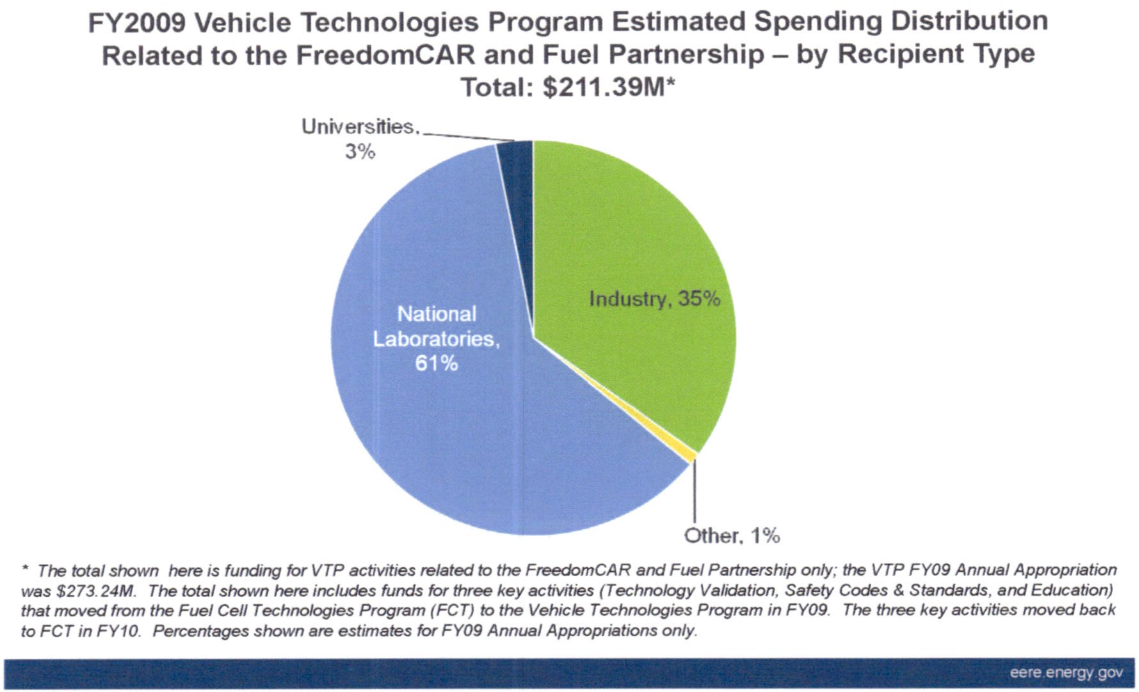

FIGURE 5-2 Distribution of funding from the Vehicle Technologies Program for FY 2009.
SOURCE: Provided to the committee by DOE, April 23, 2010.

TABLE 5-3 Estimate of DOE's Congressional Budget Request for FY 2011 FreedomCAR and Fuel Partnership (FCFP) Activities

FY 2011 Budget Structure	FY 2011 Request ($000)[a]	FCFP *Estimate*[b] ($000)
Vehicle Technologies		
Batteries and Electric Drive Technology[c]	120,637	*120,637*
Vehicle and Systems Simulation and Testing[d]	44,328	*39,127*
Advanced Combustion Engine R&D	57,600	*35,900*
Materials Technology	50,723	*38,355*
Fuels Technologies	11,000	*5,500*
Outreach, Deployment, and Analysis[e]	41,014	*3,500*
Vehicle Technologies Program Total	262,265	*243,019*
Hydrogen and Fuel Cell Technologies		
Fuel Cell Systems R&D[f]	67,000	67,000
Hydrogen Fuels R&D[g]	40,000	40,000
Systems Analysis	5,000	5,000
Market Transformation[h]	9,000	9,000
Manufacturing R&D	5,000	5,000
Technology Validation	11,000	11,000
Hydrogen and Fuel Cell Technologies Total	137,000	137,000

[a] All numbers in the Request column include SBIR/STTR.

[b] Numbers indicate *unofficial, "ballpark" estimates only* and are based solely on prior-year splits for activities supporting the FreedomCAR and Fuel Partnership, 21st Century Truck Partnership, and "other" activities. All estimates are subject to change.

[c] In the FY 2011 DOE budget request for Vehicle Technologies, the Batteries and Electric Drive Technology (BEDT) subprogram contains all of the activities of the former Hybrid Electric Systems subprogram, except for Vehicle and Systems Simulation and Testing (VSST). The proposed budget structure change gives batteries and electric/hybrid vehicles a dedicated budget line, while separating the crosscutting and non-electric/hybrid activities that are included in VSST.

[d] In the FY 2011 DOE budget request for Vehicle Technologies, the Vehicle and Systems Simulation and Testing (VSST) activity, which in prior-year budgets had been included in the former Hybrid and Electric Systems subprogram (now the Batteries and Electric Drive Technologies subprogram), has been elevated to a subprogram in order to make budget line items more transparent and meaningful.

[e] In the FY 2011 DOE budget request for Vehicle Technologies, the Technology Integration subprogram has been renamed Outreach, Deployment, and Analysis to better reflect the subprogram's activities.

[f] In the FY 2011 DOE budget request for Hydrogen and Fuel Cell Technologies, the Fuel Cell Stack Component R&D, Distributed Energy Fuel Cell Systems, Transportation Fuel Cell Systems, and Fuel Processors R&D key activities have been consolidated into a new Fuel Cell Systems R&D subprogram.

[g] In the FY 2011 DOE budget request for Hydrogen and Fuel Cell Technologies, Hydrogen Fuel R&D encompasses R&D for fuel-cell-compatible fuel production, delivery, and storage.

[h] In the FY 2011 DOE budget request for Hydrogen and Fuel Cell Technologies, a structure change consolidates the previous Safety and Codes and Standards and Education activities with early-market activities in the Market Transformation subprogram, although funding for educational activities is deferred in FY 2011.

SOURCE: Submitted to the committee by the DOE, February 2, 2010.

not in the Partnership but are supportive, so their loss would probably slow significantly, but not directly jeopardize, the existing Partnership programs.

In summary, although the committee understands the continued shift in the Partnership's funding and focus away from HFCVs and toward PHEVs and BEVs, the committee believes that in order to enable the fundamental "commercialization decision in 2015," future hydrogen-related funding should, at a minimum, be maintained at the current level. Furthermore, the committee's charter does not extend to ARRA and ARPA-E oversight, but it is very apparent, and of some concern, that the overwhelming majority of those relevant activities are devoted to only one of three possible pathways to achieving the nation's ultimate transportation energy goals.

Finally, in prior reports, the committee has expressed concern over the inclusion of congressionally directed activities (also knows as "earmarks") in DOE funding authorization for the Partnership. It is worth noting that in the FY 2010 appropriations for the DOE, Congress included many congressionally directed projects for the Office of Energy Efficiency and Renewable Energy totaling about $292 million. The DOE tentatively identified approximately $40 million of congressionally directed projects as part of the Fuel Cell Technologies or Vehicle Technologies Programs. However, Congress also appropriated additional funds specifically to support the earmarks and, as a result, the DOE-managed Partnership funding was not affected. The committee regards this as a very positive development.

CONCLUDING COMMENTS

Overall, the goals and operations of the FreedomCAR program, and then the FreedomCAR and Fuel Partnership, have changed somewhat since the creation of the FreedomCAR program in January 2002, due primarily to the balance between nearer-term and longer-term activities shifting more toward nearer-term activities. Especially considering the current economic and political issues associated with massive petroleum imports, as well as the change in administration with understandably different priorities, the committee finds such a shift to be appropriate. However, the committee also believes that an important function of the government is to support longer-term high-risk, potentially high-payoff technology developments, such as vehicle fuel cells and hydrogen as a vehicle fuel, and as such that these activities should be continued. The serious challenges that the nation faces with fuel supplies and the environment will not likely be solved with short-term solutions alone.

In terms of achievements and barriers, there has been considerable progress in most areas between Phases 2 and 3 of the National Research Council reviews just as there was between Phases 1 and 2. In some portions of the program, such as efforts toward projected fuel cell cost reductions, the results since the program began in 2002 have been very significant. In other areas, such as efforts to find onboard hydrogen storage technologies more viable than compressed-gas tanks, solutions are still elusive. In a few cases, such as materials to allow important vehicle weight reductions, technologies are lesser barriers than costs. Indeed in virtually every area being pursued within the program, projected costs continue to be major issues.

From an organizational standpoint, the committee believes that the FreedomCAR and Fuel Partnership government/industry collaboration is working well and that such precompetitive collaborations should be continued. Evidence that it is working well can be seen in the activities by and accomplishments of the technical teams, which is the heart of the Partnership. Cost and performance targets have been established in every technical area, and steady progress is noted in essentially all of them. There is, however, at least one major failing of the Partnership, the effectiveness of the Executive Steering Group (ESG). This group, composed of vice-presidential-level executives from each of the Partnership companies and the Assistant Secretary of Energy for EERE, has not met in more than 2 years. This is a problem area that should be corrected. The government involvement seems appropriate and well managed, and the government/industry technical teams have worked well to develop reasonable technical and cost goals, which have been updated when necessary. Overall, the FreedomCAR and Fuel Partnership is a very ambitious program; there is a good chance that, within the life of the program, some of the key targets will never be met. However, this program with its goals and long-term collaborations is, in the opinion of the committee, very much in the nation's interest.

REFERENCES

DOE (Department of Energy). 2004. *Hydrogen Posture Plan: An Integrated Research, Development and Demonstration Plan*. Washington, D.C.: U.S. Department of Energy. Available on the Web at <http://www.eere.energy.gov/hydrogenandfuelcells/pdfs/hydrogen_posture_plan.pdf>.

DOE. 2005. *Multi-Year Research, Development and Demonstration Plan: Planned Program Activities for 2005-2015*. Section 2.3, Domestic Resources for Hydrogen Production. Washington, D.C.: Office of Energy Efficiency and Renewable Energy. Available on the Web at <http://www1.eere.energy.gov/hydrogenandfuelcells/mypp/>.

DOE. 2008. "FreedomCAR and Fuel Partnership 2008 Highlights of Technical Accomplishments." Available on the Web at <http://www1.eere.energy.gov/vehiclesandfuels/pdfs/program/2008_fcfp_accomplishments_rpt.pdf>.

Ge, H.-W., Y. Shi, R.D. Reitz, and W. Willems. 2010. "Engine Development Using Multi-Dimensional CFD and Computer Optimization." SAE Paper 2010-01-0360.

NRC (National Research Council). 2005. *Review of the Research Program of the FreedomCAR and Fuel Partnership, First Report*. Washington, D.C.: The National Academies Press.

NRC. 2008. *Review of the Research Program of the FreedomCAR and Fuel Partnership, Second Report*. Washington, D.C.: The National Academies Press.

NRC/NAE (National Research Council/National Academy of Engineering). 2004. *The Hydrogen Economy: Opportunities, Costs, Barriers, and R&D Needs*. Washington, D.C.: The National Academies Press.

Satyapal, S., and P. Davis. 2009. Changes in Objectives of the Partnership. Presentation to the Committee on Review of the Research Program of the FreedomCAR and Fuel Partnership, Phase 3, April 27, Washington, D.C.

Appendixes

Appendix A

Biographical Sketches of Committee Members

Vernon P. Roan, *Chair,* is retired director of the Center for Advanced Studies in Engineering and a professor of mechanical engineering at the University of Florida, where he has been a faculty member for more than 30 years. Since 1994, he has also been the director of the University of Florida Fuel cell Research and Training Laboratory. Previously he was a senior design engineer with Pratt and Whitney Aircraft. Dr. Roan has more than 25 years of research and development experience as well as modeling and simulation experience for a fuel cell bus program. He worked as a consultant to Pratt and Whitney on advanced gas-turbine propulsion systems until his 2007 retirement from that position. His research at the University of Florida has involved both spark-ignition and diesel engines operating with many alternative fuels and advanced concepts. With groups of engineering students, he designed and built a 20-passenger diesel-electric bus for the Florida Department of Transportation and a hybrid–electric urban car using an internal-combustion engine and lead-acid batteries. He has been a consultant to the Jet Propulsion Laboratory, monitoring their electric and hybrid vehicle programs. He has organized and chaired two national meetings on advanced vehicle technologies and a national seminar on the development of fuel-cell-powered automobiles and has published numerous technical papers on innovative propulsion systems. He was one of four members of the Fuel Cell Technical Advisory Panel of the California Air Resources Board (CARB), which issued a report in May 1998 regarding the status and outlook for fuel cells for transportation applications. He also served as one of five members of CARB's Zero Emission Vehicle Expert Panel, which issued a report on the *Status and Prospects for Zero Emission Vehicle Technology* in April 2007. He has served on numerous National Research Council (NRC) committees, including the Committee on Review of the FreedomCAR and Fuel Research Program,

Phase 1 and Phase 2, and the prior Committee to Review the Research Program of the Partnership for a New Generation of Vehicles. Dr. Roan received his B.S. in aeronautical engineering and his M.S. in engineering from the University of Florida and a Ph.D. in engineering from the University of Illinois.

Deborah Lynn Bleviss is an independent consultant focused on studies and analyses on sustainable energy and transportation both internationally and domestically. Recent studies that she has worked on have included options for biofuels development, and an evaluation of the barriers and potential solutions to clean-energy financing, in Latin America and the Caribbean. Her previous positions include partner, the BBG Group; program manager, Sustainable Markets for Sustainable Energy, Inter-American Development Bank; consultant to the Department of Energy's Assistant Secretary for Energy Efficiency and Renewable Energy; executive director and president of the board of directors, International Institute for Energy Conservation; and associate director for Energy and Environment, Federation of American Scientists. She has extensive experience in the impact of transportation on the environment, and with strategies, both technical and policy, related to the development and deployment of transportation systems more conducive to sustainability. She has served on numerous advisory councils and committees and was the lead author, Second Assessment Report of the Intergovernmental Panel on Climate Change, 1994-1995. She has written extensively on transportation, vehicles, energy, and the environment. She has a B.S. in physics from the University of California, Los Angeles.

David L. Bodde serves as a professor and senior fellow at Clemson University. Prior to joining Clemson University, Dr. Bodde held the Charles N. Kimball Chair in Technology and Innovation at the University of Missouri in Kansas City. Dr. Bodde serves on the board of directors of several energy and technology companies, including Great Plains Energy and the Commerce Funds. His executive experience includes vice president, Midwest Research Institute; president, MRI Ventures; assistant director of the Congressional Budget Office; and deputy assistant secretary in the U.S. Department of Energy. He has served as a member of the NRC's Board on Energy and Environmental Systems, the Committee on Alternatives and Strategies for Future Hydrogen Production and Use, and the Committee on Assessment of Resource Needs for Fuel Cell and Hydrogen Technologies. He was once a soldier and served in the U.S. Army in Vietnam. He has a doctorate in business administration from Harvard University, M.S. degrees in nuclear engineering (1972) and management (1973), and a B.S. from the United States Military Academy.

Kathryn Bullock is the president and founder of Coolohm, Inc., which is a technical consulting company that specializes in direct current (dc) power sources such as batteries, capacitors, and fuel cells and their application in electronic systems.

She is also an adjunct faculty member at Villanova University, where she teaches a course on Electrochemical Power Sources, including fuel cells, batteries, and capacitors and their application in dc power systems. Her previous positions include vice president, C&D Technologies, Inc., where she was responsible for the development of new battery products and new product applications such as solar energy and fuel cell systems and for providing technical leadership and support to executive and board members; development manager, power sources, Medtronic, Inc. Promeon Division; technical manager, Batteries and Purchased Products, Lucent Technologies, Bell Laboratories (Mesquite, Texas); and manager, Chemical Research Department, and senior electrochemist, Electrochemical Research Department, Johnson Controls, Inc. She has extensive research and development and manufacturing experience in electrochemical devices, including batteries and capacitors. She has a Ph.D. in physical chemistry and an M.S. in chemistry from Northwestern University and a B.A. in English from Colorado University.

Harry E. Cook (NAE) is professor emeritus, Department of General Engineering, University of Illinois. He is a recipient of the Robert Lansing Hardy Medal and the Teetor Award. He has also received awards from the American Institute of Mining and Metallurgical Engineers, is a fellow of the Society of Automotive Engineers, and a fellow of the American Society of Metals. His career in the automotive industry began at the Ford Motor Company as a senior research engineer and culminated with his position as the director of automotive research with Chrysler Motors. Dr. Cook was also a professor with the University of Illinois in the Department of Mechanical and Industrial Engineering and director of the Manufacturing Research Center. His research experience includes phase transformations, friction and wear, automotive product development, value engineering, and competitiveness. He received his Ph.D. in materials science from Northwestern University, and an M.S. and B.S. in metallurgical engineering from Case Western Reserve University.

Glenn A. Eisman is a principal partner, Eisman Technology Consultants, LLC, a managing partner at H2Pump LLC, and an adjunct professor at Rensselaer Polytechnic Institute in materials science and engineering (Troy, N.Y.), and at the Graduate College of Engineering at Union University (Schenectady, N.Y.). His previous positions include chief technology officer, Plug Power, Inc.; technical leader, Advanced Materials Program, Central Research and New Businesses, The Dow Chemical Company; project leader, Discovery Research R&D and product development of fuel cells, hydrogen technologies, electrochemical engineering, physical and inorganic solid-state chemistry, and new technology commercialization and business development. He received the Inventor of the Year Award, from the Dow Chemical Co. (1993) and is a member of the Electrochemical Society. He received a B.S. in chemistry from Temple University and a Ph.D. in physical inorganic chemistry from Northeastern University. He has published more than 20 technical papers and has been awarded more than 20 U.S. patents.

W. Robert Epperly is an independent consultant. From 1994 to 1997, he was president of Catalytica Advanced Technologies, Inc., a company developing new catalytic technologies for the petroleum and chemical industries. Prior to joining Catalytica, he was general manager of Exxon Corporate Research and earlier was director of the Exxon Fuels Research Laboratory. After leaving Exxon, he was chief executive officer of Fuel Tech N.V., a company developing new combustion and air pollution control technology. Mr. Epperly has authored or coauthored more than 50 publications on technical and managerial topics, including two books, and has 38 U.S. patents. He has extensive experience in the conversion of fossil feedstocks to alternative fuels such as gases and liquids, fuels, catalysis, air pollution control, and R&D management. He received an M.S. degree in chemical engineering from the Virginia Polytechnic Institute and State University.

William D. Ernst is an independent consultant. He retired from Plug Power, Inc., as vice president and chief scientist. There he was responsible for proton exchange membrane fuel cell technology assessment and advanced development, as well as technical initiatives within the government sector. Most recently, he investigated the applicability of solid oxide fuel cell technology to various continuous power applications. Prior to joining Plug Power, Dr. Ernst was business area manager of the Technology Division at Mechanical Technology Incorporated (MTI), where he was responsible for the management and development of the fuel cell, hybrid electric vehicle, and flywheel business. His other positions at MTI included business development manager, manager for the Kinematic and Advanced Power System Programs, and program manager for the Automotive Stirling Engine program. Previously, Dr. Ernst founded a consulting/engineering business and held positions with Huyck Corporation and Ling Tempco Vought. He is the author of more than 100 technical reports and papers on subjects including proton exchange membrane fuel cell technology development and application and non-Newtonian fluid dynamics. Dr. Ernst is the recipient of the 1998 Partnership for the Next Generation of Vehicles Award. He received a B.S. in engineering from Tufts University, an M.S. in engineering from the Massachusetts Institute of Technology, and a Ph.D. in aeronautical engineering from the Rensselaer Polytechnic Institute.

David E. Foster is a professor of mechanical engineering, University of Wisconsin, Madison, and former director of the Engine Research Center, which has won two center of excellence competitions for engine research and has extensive facilities for research on internal combustion engines. A member of the faculty at the University of Wisconsin since he completed his Ph.D., Dr. Foster teaches and conducts research in thermodynamics, fluid mechanics, internal combustion engines, and emission-formation processes. His work has focused specifically on perfecting the application of optical diagnostics in engine systems and the incorporation of simplified or phenomenological models of emission-formation processes into engineering simulations. He has published more than 60 technical articles in this field throughout the world and for leading societies in this country. He is a recipient of the Ralph R. Teetor Award, the Forest R. McFarland Award, and the Lloyd L. Withrow Distinguished Speaker Award of the Society of Automotive Engineers (SAE) and is an SAE Fellow. He has served on a number of NRC committees, including the Committee to Review the Research Program of the Partnership for a New Generation of Vehicles. He is a registered professional engineer in the State of Wisconsin and has won departmental, engineering society, and university awards for his classroom teaching. He received a B.S. and M.S. in mechanical engineering from the University of Wisconsin and a Ph.D. in mechanical engineering from the Massachusetts Institute of Technology.

Gerald Gabrielse (NAS) is Leverett Professor of Physics at Harvard University. His previous positions include assistant and associate professor, University of Washington-Seattle, and chair of the Harvard Physics Department. His physics research focuses on making the most accurate measurements of the electron magnetic moment and the fine structure constant, and on the precise laser spectroscopy of helium. Professor Gabrielse also leads the International Antihydrogen TRAP (ATRAP) Collaboration, whose goal is accurate laser spectroscopy with trapped anti-hydrogen atoms. His many awards and prizes include fellow of the American Physical Society, Davisson-Germer prize of the American Physical Society, the Humboldt Research Award (Germany, 2005) and the Tomassoni Award (Italy, 2008). Harvard University awarded him both its George Ledlie Research Prize and its Levenson Teaching Prize. His hundreds of outside lectures include a Källén Lecture (Sweden), a Poincaré Lecture (France), a Faraday Lecture (Cambridge, U.K.), a Schrodinger Lecture (Austria), a Zachariasen Lecture (University of Chicago), and a Rosenthal Lecture (Yale). He is a member of the National Academy of Sciences. He has a B.S. from Calvin College, and an M.S. and Ph.D. in physics from the University of Chicago.

Linos Jacovides recently retired as director, Delphi Research Laboratories, a position that he held from 1998 to 2007. Dr. Jacovides joined General Motors (GM) Research and Development in 1967 and became department head of electrical engineering in 1985. His areas of research were the interactions between power electronics and electrical machines in electric vehicles and locomotives. He later transitioned to Delphi with a group of researchers from GM to set up the Delphi Research Laboratories. He is a fellow of the Institute of Electrical and Electronics Engineers (IEEE) and was president of the Industry Applications Society of IEEE in 1990. He received a B.S. degree in electrical engineering and an M.S. in machine theory from the University of Glasgow, Scotland, in 1961 and 1962, respectively. He received his Ph.D. in generator control systems from the Imperial College, University of London, in 1965.

Harold H. Kung is a professor of chemical engineering and director of the Center for Energy Efficient Transportation at Northwestern University. His areas of research include surface chemistry, catalysis, and chemical reaction engineering. His professional experience includes work as a research chemist at E.I. du Pont de Nemours & Co., Inc. He is a recipient of the P.H. Emmett Award and the Robert Burwell Lectureship Award from the North American Catalysis Society, the Herman Pines Award of the Chicago Catalysis Club, the Japanese Society for the Promotion of Science Fellowship, the John McClanahan Henske Distinguished Lectureship of Yale University, and the Olaf A. Hougen Professorship at the University of Wisconsin, Madison. He has a Ph.D. in chemistry from Northwestern University.

Christopher L. Magee (NAE) is a professor, Engineering Systems Division, Massachusetts Institute of Technology (MIT), and director, Center for Innovation in Product Development. Prior to joining MIT, he held a number of positions at Ford Motor Company, including director, Vehicle Systems Engineering; director, Advanced Vehicle Engineering; manager, Materials Science Department; senior research scientist, Metallurgy Department; and executive director, Programs and Advanced Engineering, with global responsibility for all major technically deep areas involved in Ford's Product Development Organization. He has expertise in such areas as phase transformations, plastic deformation, materials strength, large-scale collapse of engineering structures, product development, automotive design, value engineering, and simultaneous manufacturing/product engineering. He has made important contributions to the understanding of the transformation, structure, and strength of ferrous materials and to lightweight materials development and implementation; he pioneered experimental work on high-rate structural collapse aimed at vehicle crashworthiness; and he adapted systems engineering to the modern automotive design process. He was elected to the National Academy of Engineering for contributions to advanced vehicle development, was a Ford Technical Fellow (1996), and is a fellow of the American Society for Materials. He has a B.S., an M.S., and a Ph.D. in metallurgy and materials science from the Carnegie Institute of Technology (now Carnegie Mellon University) and an M.B.A. from Michigan State University.

Gene Nemanich is the retired vice president of Hydrogen Systems for Chevron Technology Ventures where he was responsible for hydrogen supply and for developing and commercializing new hydrogen technologies. He has 32 years of experience with integrated oil companies, including Exxon, Cities Service, Texaco, and Chevron. He has also worked in the areas of refining, clean coal technology, oil supply and trading, and research leading to the development of new hydrogen systems. Mr. Nemanich represented Texaco in the California Fuel Cell Partnership in 2000-2001 and was a director of Texaco Ovonic Hydrogen Systems LLC, a joint venture with Energy Conversion Devices to commercialize metal hydride hydrogen storage systems. He was one of seven industry leaders who helped prepare the DOE-sponsored Hydrogen Roadmap, and he has served as chair of the National Hydrogen Association. He recently served on the NRC Committee on Assessment of Resource Needs for Fuel Cell and Hydrogen Technologies. He has a B.S. in chemical engineering from University of Illinois and an M.B.A. from the University of Houston.

Bernard Robertson (NAE) is the president of BIR1, LLC, an engineering consultancy specializing in transportation and energy matters that he founded in January 2004, upon his retirement from DaimlerChrysler Corporation. During the latter part of his 38-year career in the automotive industry, Mr. Robertson was elected an officer of Chrysler Corporation in February 1992. He was appointed senior vice president coincident with the merger of Chrysler Corporation and Daimler-Benz AG in November 1998, and was named senior vice president of engineering technologies and regulatory affairs in January 2001. In his last position, he led the Liberty and Technical Affairs Research Group, Advanced Technology Management and FreedomCAR activities, and hybrid electric, battery electric, fuel cell, and military vehicle development. In addition, he was responsible for regulatory analysis and compliance for safety and emissions. Mr. Robertson holds an M.B.A. degree from Michigan State University, a master's degree in automotive engineering from the Chrysler Institute, and a master's degree in mechanical sciences from Cambridge University, England. He is a member of the National Academy of Engineering, a fellow of the Institute of Mechanical Engineers (U.K.), a chartered engineer (U.K.), and a fellow of the Society of Automotive Engineers.

R. Rhoads Stephenson is currently a technology consultant. Previously, he held a number of positions at the Jet Propulsion Laboratory (JPL), the National Highway Traffic Safety Administration (NHTSA), and Martin Marietta Corporation. At JPL, these included deputy director and acting director, Technology and Applications Programs; manager, Electronics and Control Division; deputy manager, Control and Energy Conversion Division; and manager of Systems Analysis Section. He also served as associate administrator for research and development, NHTSA and while at Martin Marietta Corporation worked on energy conversion devices for space power. He has been a consultant to the Motor Vehicle Fire Research Institute, has been providing peer reviews of automotive safety issues, and has recently published a number of papers on crash-induced fire safety issues with motor vehicles, including hydrogen-fueled vehicles. He brings extensive expertise in vehicle safety analysis, advanced technology systems, energy conversion technologies, and energy and environmental analysis. He has a B.S., M.S., and Ph.D. in mechanical engineering from Carnegie Mellon University.

Kathleen C. Taylor (NAE) is retired director of the Materials and Processes Laboratory at General Motors Research and Development and Planning Center in Warren, Michigan. Dr. Taylor was simultaneously chief scientist for General Motors of Canada, Ltd. in Oshawa, Ontario. Earlier Dr. Taylor was department head for physics and physical chemistry and department head for environmental sciences. Currently, Dr. Taylor serves on the DOE Hydrogen Technology Advisory Committee, the Transportation Research Board Committee for a Study of Potential Energy Savings and Greenhouse Gas Reduction from Transportation, the DOE Basic Energy Sciences Advisory Committee, the DOE Materials Forum, and the Advisory Committee for Columbia University Center for Electron Transport in Molecular Nanostructures. Dr. Taylor was awarded the Garvan Medal from the American Chemical Society. She is a member of the National Academy of Engineering, the American Academy of Arts and Sciences, and the Indian National Academy of Engineering and a fellow of SAE International and the American Association for the Advancement of Science. She was the president of the Materials Research Society and chair of the board of directors of the Gordon Research Conferences. She has expertise in R&D management, fuel cells, batteries, catalysis, exhaust emission control and automotive materials. She received an A.B. in chemistry from Douglass College and a Ph.D. in physical chemistry from Northwestern University.

Brijesh Vyas is a distinguished member of the technical staff at LGS Innovations, LLC. Previously he was a member of the Nanotechnology and Integrated Photonic Research Departments at Bell Labs, Murray Hill, N.J., responsible for advanced materials and processes for microelectromechanical systems and photonic devices. He was also the technical manager of the Energy Conversion Technology Group responsible for research on advanced materials and technologies for energy storage systems. He has led efforts to develop various rechargeable batteries and related energy conversion technologies for a variety of telecommunications applications. He was formerly at the Brookhaven National Laboratory and has been a guest professor at the Technical University of Denmark in Copenhagen investigating the corrosion and erosion of metals. He received the Sam Tour Award from the American Society of Materials and Testing. His areas of expertise include materials science, electrochemistry, and corrosion. He served on the NRC Committee to Review the U.S. Advanced Battery Consortium's electric vehicle battery R&D project selection process. He received a bachelor's degree in metallurgical engineering from the Indian Institute of Technology in Bombay and a Ph.D. in materials science from the State University of New York, Stony Brook.

Eric Williams is research director, Center for Earth Systems Engineering and Management and assistant professor, Department of Civil, Environmental and Sustainable Engineering, School of Sustainability at Arizona State University. His research interests include industrial ecology, life-cycle assessment, information technology, and energy systems. His best-known work addresses the environmental assessment and management of information technology hardware. Dr. Williams also investigates energy topics such as long-term, second-law efficiency trends and the effects of development and urbanization on energy demand in industrializing nations. He has worked in the areas of hybrid life-cycle assessment (which combines process and economic input-output techniques), uncertainty analysis in industrial ecology, and the sector-level forecasting of technological change and growth. His areas of expertise include industrial ecology, life-cycle assessment, and the macro-assessment of energy supply and demand. He received his Ph.D. in physics from the State University of New York.

Appendix B

Committee's Interim Letter Report

THE NATIONAL ACADEMIES
Advisers to the Nation on Science, Engineering, and Medicine

National Research Council
Board on Energy and Environmental Systems

500 Fifth Street, NW
Washington, DC 20001
Phone: 202 334 3344
Fax: 202 334 2019

July 10, 2009

The Honorable Steven Chu
Secretary
U.S. Department of Energy
1000 Independence Avenue, SW
Washington, DC 20585

The Honorable Cathy Zoi
Assistant Secretary of Energy Efficiency and Renewable Energy
U.S. Department of Energy
1000 Independence Avenue, SW
Washington, DC 20585

Dear Secretary Chu and Assistant Secretary Zoi:

This National Research Council (NRC) letter report was prepared by the Committee on Review of the Research Program of the FreedomCAR and Fuel Partnership, Phase 3 (see Attachment I), in response to a request from the U.S. Department of Energy (DOE) (see Attachment II for the statement of task). It addresses one part of its statement of task, namely, to broadly review the strategy and structure of the FreedomCAR and Fuel Partnership (hereafter referred to as the Partnership). Attachment III lists the presentations to the committee at its April 27, 2009 meeting. The committee welcomes the chance to offer the benefit of its experience and expertise in providing some suggestions and guidance to the Partnership as it addresses future challenges and reviews its goals, strategy, organization, and priorities.

The committee recognizes and agrees with the new Administration's focus on nearer-term technologies. However, it also emphasizes the need for continued investment in longer-term, higher-risk, higher-payoff vehicle technologies that could be highly transformational with regard to reduced use of petroleum and reduced emissions. Such technologies include advanced batteries, technologies for hydrogen storage, and hydrogen/fuel cells. The committee has also concluded that for researchers, contractors, and investors to be willing to make long-term commitments to these and other potentially important developing technologies, a consistent year-to-year level of support must be provided.

The committee has further concluded that, given increasing concerns about greenhouse gas (GHG) emissions and world climate change, the Partnership should incorporate in its planning a broader-scope, "cradle-to-grave" analysis rather than a "well-to-wheels" approach, to better consider total emissions and the full environmental impact of using various fuels and technologies. In addition, the Partnership should consider broadening the scope of technical approaches being considered within each of what the committee considers to be the three major fuel and vehicle pathways—biofuels/internal combustion engine (ICE) vehicles, plug-in electric vehicles (PHEVs)/battery-electric vehicles (BEVs), and hydrogen-fueled fuel cell vehicles.

Finally, the committee concluded that several measures should be considered by DOE to assist in implementing these suggestions. One is to provide temporary reductions in cost-share requirements to ease the burden on prospective researchers. Otherwise, there could be a significant number of potential worthy contributors who cannot afford the matching funds. Another implementation suggestion, occasioned by the obvious financial problems of the automotive companies (OEMs), is to consider providing direct funding to them to help keep important in-house research programs active. Other suggestions are included in the balance of the report.

INTRODUCTION

The Partnership,[1] as it currently exists, can be described as a focused research and technology development program that emphasizes high-risk, high-payoff technologies believed to be essential for a transition to vastly different light-duty passenger vehicles. "Vastly different" means vehicles that, according to the Partnership's original long-term goals, differ from existing light-duty vehicles (LDVs) in that they include the possibility of a full spectrum of vehicles that can operate without petroleum and free of harmful emissions while sustaining the driving public's freedom of mobility and freedom of vehicle choice. The needed research has been directed and supported by a collaboration among the U.S. government (especially DOE), the United States Council for Automotive Research (USCAR; its members are Chrysler LLC, Ford Motor Company, and General Motors Corporation), five key energy companies (BP America, Chevron Corporation, ConocoPhillips, ExxonMobil Corporation, and Shell Hydrogen [U.S.]), and more recently two major utility companies (Southern California Edison and DTE Energy) (DOE, 2006, 2009a). The Partnership has established, and periodically reviews, a roadmap with research milestones against which to measure progress in

[1] As described in DOE (2006), the "Partnership" is not a legal entity, and it is not intended that the "partners" have the responsibilities or rights of legal partners. Rather, the terms "Partnership" and "partners" are used in an informal sense to denote participants working together toward the stated goals of the group.

moving toward long-term goals. The long-term goals have focused on hydrogen/fuel cell vehicles. (For further information see Attachment IV and previous NRC [2005, 2008a] reports.)

Two reports by the NRC Committee on Review of the Research Program of the FreedomCAR and Fuel Partnership have assessed the structure and management of the Partnership as well as the nature, adequacy, and progress of the research activities (NRC, 2005, 2008a). A third report, based on Partnership activities and progress following publication of the Phase 2 report (NRC, 2008a), is planned to be issued during this third review. However, a number of recent changes in policy as well as technology advancements, described below, will influence the long-term goals of the Partnership as well as the paths to achieving them. In response to a request by DOE that the committee start its Phase 3 work by writing a letter report on the effects of these events and suggesting corresponding changes in the program, work on the Phase 3 report was temporarily delayed. This brief interim letter report is an attempt by the committee to offer constructive suggestions for possible changes to the existing Partnership program, especially its goals and strategy.

GOALS

The long-term goal of the Partnership has been to enable the transition to a transportation system "that uses sustainable energy resources and produces minimal criteria[2] or net carbon emissions on a life cycle or well-to-wheel basis" (DOE, 2006, p. iii).

Achievement of the sustainability goal will also contribute to reducing U.S. dependence on petroleum, another important national objective. A recent NRC report (NRC, 2008b) concluded that hydrogen-fueled fuel cell vehicles offer greater long-term potential for reducing U.S. dependence on imported petroleum and reducing carbon emissions significantly by 2050 than would relying only on fuel economy improvements (e.g., through engine efficiency improvements) and increased use of biofuels.

The Partnership plan envisions a pathway starting with more fuel-efficient ICEs and hybrid-electric vehicles (HEVs), including PHEVs, potential use of all-electric drive vehicles (BEVs), and, ultimately, hydrogen-fueled fuel-cell vehicles concurrent with the addition of an infrastructure for supplying hydrogen fuel. The lightweight materials program will continue to be an integral part of the efforts to improve vehicle fuel economy. It is expected that the Partnership plan will be modified to be more consistent with priorities of the new Administration outlined as major points made in April 2009 presentations to this committee. Among the new Administration's goals and priorities, which will obviously affect the Partnership, are the following as listed by Satyapal and Davis (2009a):

[2] Criteria emissions refer to those that are regulated by law.

New Energy for America Presented by Obama-Biden Administration

- Help create 5 million new jobs by strategically investing $150 billion over the next 10 years to catalyze private efforts to build a clean energy future.
- Within 10 years save more oil than we currently import from the Middle East and Venezuela combined.
- Put 1 million plug-in hybrid cars—cars that can get up to 150 miles per gallon—on the road by 2015, cars that we will work to make sure are built here in America.
- Ensure that 10 percent of our electricity comes from renewable sources by 2012, and 25 percent by 2025.
- Implement an economy-wide, cap-and-trade program to reduce greenhouse gas emissions by 80 percent by 2050.

Energy to Secure America's Future:
President's National Objectives for DOE

- Quickly implement the economic recovery package—Create millions of new Green jobs and lay the foundation for the future.
- Restore science leadership—Strengthen America's role as the world leader in science and technology.
- Reduce GHG emissions—Drive emissions 20 percent below 1990 levels by 2020.
- Enhance energy security—Save more oil than the U.S. currently imports from the Middle East and Venezuela combined, within 10 years.
- Enhance nuclear security—Strengthen non-proliferation activities. Reduce global stockpiles of nuclear weapons, and maintain safety and reliability of the U.S. stockpile.

Secretary of Energy Chu's Priorities

- Focus on transformational science:
  —Connect basic and applied sciences
  —Embrace a degree of risk-taking
  —Integrate lab, university and industry activities
- Collaborate universally:
  — Build research networks and global partners
- Demonstrate next-generation energy technologies:
  — Batteries and other storage systems
- Drive step-change energy efficiency:
  — Novel models for collaboration and [use of] intellectual property (IP) for commercialization of energy-efficient technologies

- Reduce vehicle energy demand:
  — Improve internal combustion engines and develop batteries for vehicle electrification
- Build an efficient, smart network:
  — Smart meters/smart grid (and vehicle interface)
- Coordinate and share research globally

Technically, these priorities and goals are expected by the committee to translate into research, development, demonstration, and deployment (RDD&D) of efficient low-carbon transportation technologies and RDD&D of PHEV technology/vehicle electrification. From a programmatic standpoint, they suggest a greater emphasis on manufacturing, production, and commercialization.

For the Partnership, the net effect will be determined partially by the DOE's FY2010 budget request (Chu, 2009) that "cuts less effective programs so we can invest in our economic future." One of the examples mentioned by Secretary Chu is "moving away from funding vehicular hydrogen fuel cells to technologies with more immediate promise." This approach is indeed reflected in the DOE FY2010 budget request to Congress for $0 for hydrogen technologies as compared to $168,960,000 for FY2009 (DOE, 2009b). On the other hand, the vehicle technologies request is up from $273 million to $333 million in addition to requests for biofuel, solar, and wind technologies, which are also up. Further indications of priorities for vehicle electrification are that, of DOE's Office of Energy Efficiency and Renewable Energy's (EERE's) $4.5 billion American Recovery and Reinvestment Act 2009 (ARRA) funds, $2 billion are proposed for advanced battery manufacturing, $400 million for transportation electrification, and $300 million for the alternative-fueled vehicles pilot grant program (Satyapal and Davis, 2009b).

In apparent support of these priorities, in DOE's Office of Electricity Delivery and Energy Reliability, about $3.4 billion of its $4.5 billion ARRA funds will go to the "Smart Grid Investment Program (Energy Independence and Security Act [EISA] 1306)" and another $700 million to "Smart Grid Regional and Energy Storage Demos" (DOE, 2009b). This distribution is consistent with increased emphasis on PHEVs and BEVs, since the need for Smart Grid technologies increases as the number of such vehicles in the fleet is increased.

Given that the previous administration's priorities were more focused on hydrogen and fuel-cell-powered vehicles for the long term while envisioning that advanced ICEs and HEVs would provide transition technologies, the DOE involvement in the Partnership is virtually certain to change. Recognizing that Partnership changes were likely, the statement of task for the NRC's third review by this committee includes the following: " . . . in examining the Partnership strategy, and given the changes that may take place with the New Administration, the committee at its first full committee meeting will address potential changes in the program strategy and program structure. The committee will write a short interim letter report with suggestions and recommendations on program strategy and structure . . . . " (see Attachment II). That portion of the statement of task is the basis for this letter report. Subsequent meetings by this committee will allow more in-depth review of the progress in the various technology areas and reporting of the Partnership activities since Phase 2.

EVENTS AND DRIVERS FOR POTENTIAL CHANGES IN THE PARTNERSHIP

A number of events since the last NRC review (NRC, 2008a) could have a significant impact on both the technological and the societal goals of the Partnership. Among these, in addition to changes in Administration/DOE priorities, are advances in battery technologies, the continued evolution of biofuels, promising basic ICE research on fuels and combustion processes, increased emphasis on reducing GHG emissions, important advances in learning from vehicle and fuel cell demonstration programs, other basic research programs, and economic issues for both the auto industry and the nation as a whole. Some of these are discussed briefly below.

Because the committee's in-depth review of the Partnership activities since Phase 2 will begin very soon, hydrogen/fuel cell technologies are not discussed to any significant extent in this letter report. However, the committee is concerned about the impact of severely scaling back the DOE hydrogen/fuel cell vehicle programs. It is not yet clear that the hydrogen/fuel cell approach (or for that matter advanced ICEs/biomass, or PHEVs/BEVs) can or cannot meet reasonable emission and driving-range requirements while also being affordable to purchase and operate. Recent fuel cell lifetime and durability improvements are encouraging, as are projected lower costs. Further, even though demonstration hydrogen/fuel cell vehicles are showing safe operation at ever-increasing driving ranges with compressed hydrogen gas storage, the existing DOE hydrogen storage centers of excellence, in the committee's view, are likely to provide the best opportunity for finding better solutions, if they exist. An in-depth evaluation of the hydrogen/fuel cell option will be part of the current review and the committee's final report.

The Evolution of New Relevant Technologies

Batteries

Cooperative alternative energy automotive programs were greatly influenced by the formation and activities of the United States Advanced Battery Consortium (USABC). The cooperative efforts continued into the Partnership for a New Generation of Vehicles (PNGV) program and later into the FreedomCAR and then the Partnership. An NRC review of this program in 1998 concluded that both nickel- and lithium-based batteries had the potential to meet the automobile industry goals, except for the costs (NRC, 1998).

The evolution of lithium-ion batteries as candidates for HEVs, PHEVs, and BEVs is known and well documented. Several OEMs have already indicated their strong interest in Li-ion, especially for PHEVs and extended-range electric vehicle (EREV)[3] applications, and efforts are underway to construct several new manufacturing facilities in the United States to produce these batteries. Although much more work needs to be done to meet all of the requirements, significant advances have been made in safety, performance increases, and cost reductions for several Li-ion chemistries. While the advances appear to be nearing adequacy for PHEVs and EREVs, this is probably not the case for BEVs, where battery cost and longer range are even more important.

The committee expects that BEVs will represent one of the important vehicle technologies in the mix of technologies for meeting the long-term goals of the Partnership. As such, even if PHEV requirements can soon be met (and certainly if they cannot), efforts to further advance Li-ion technologies are warranted, the committee suggests. It also suggests that lower-cost energy storage technologies, including other battery chemistries, advanced materials for electrochemical capacitors, and combinations of the two, be pursued for both PHEV and BEV applications.

In the PNGV many forms of energy storage and conversion technologies such as flywheels, fuel cells, and ultracapacitors, in addition to a range of batteries, were considered in an effort to meet the miles per gallon (mpg) fuel economy goal of "up to 80 mpg." That work concluded with all three of the PNGV concept cars embracing the HEV configuration using batteries and small diesel engines (NRC, 2000, 2001). In the Partnership, the current plan is to use the continued development of HEV technologies as an interim step toward a final vision of a hydrogen-fueled fuel cell vehicle.

HEVs were introduced into the marketplace about 12 years ago and today account for about 3 percent of the new-vehicle automotive market.[4] The acceptance and success of HEVs and the need to accelerate reductions in emissions have resulted in a change in the Partnership plan to include the development of technologies needed for PHEVs. PHEVs are characterized by an increase in the "battery only" range of the hybrid vehicle and a corresponding decrease in onboard fuel consumption. Further improvements in battery technologies and further reductions in cost beyond those needed for PHEVs could allow an even greater increase in the battery driving range in support of producing BEVs, which would use no onboard fuel. For HEVs, PHEVs, and BEVs the primary issues are performance, durability, and the costs of the battery system.

---

[3] General Motors refers to its "Volt"-type vehicle as an "extended-range electric vehicle (EREV)." The configuration is very similar to that of PHEVs.

[4] Note that with the recent drop in gasoline prices over the last year or so, the market share of new HEVs sold has fallen below this level. Of note is that it has taken more than a decade for this new vehicle technology to achieve a very small share of the new vehicle market sales.

ICE Emerging Technologies[5]

The successful development of biofuel production technologies such as cellulosic ethanol may offer synergistic opportunities with the advanced combustion technologies that are being researched within the Partnership. Successful technologies for the production of biofuels and advanced technologies for the use of biofuels in vehicles could impact the technology pathway and requisite timeline in which PHEVs, BEVs, and hydrogen-fueled fuel cell vehicles are introduced. For example, it may be feasible to marry the interaction between advanced fuel injection systems and in-cylinder fluid mechanics, which is currently being investigated, with the enhanced understanding and predictability of the kinetic pathways to combustion auto-ignition, also a research area within the Partnership, to capitalize on specific fuel characteristics. This enhanced understanding could point the way to the different, and perhaps "tailorable," auto-ignition characteristics of synthesized biofuels. If this were to be done successfully it could enable the development of a clean-burning, biofueled, ICE-powered hybrid vehicle. Such a vehicle could offer very efficient mobility with a minimal carbon footprint. This alternative was not a consideration prior to the initiation of the biofuels program, which was not in effect at the beginning of the Partnership.

Although the Partnership has not focused on biofuels to this point, the synergies noted in this letter report suggest that biofuels are likely to become increasingly important and may well be addressed more specifically in the future. If that were to happen, one important step would be to apply the same cradle-to-grave analysis this letter report advocates for the other technologies being addressed by the Partnership.

Changes in the Automotive Industry

In the past 12 months the world has experienced extreme volatility in the price of crude oil and the corresponding cost of fuel for the consumer. Very high fuel prices caused a rapid ramp-up in consumers' preference for fuel efficient vehicles and in their willingness to pay a premium for vehicles with high fuel economy. This preference for fuel efficient vehicles was quickly followed by a decrease in this unique demand as fuel prices came down. The industry infrastructure is not equipped to respond to such a rapid change in product mix.

In addition to the high fuel prices, the dramatic changes in the worldwide economic outlook in the last several months have also had a major impact on the automotive industry. Automobiles are a high-cost purchase, typically the second largest purchase after housing, and robust auto sales require a healthy credit market for both the dealerships and the consumer. The collapse of the credit market meant

[5] The NRC Committee on Light-Duty Vehicle Fuel Economy Technologies is investigating a variety of options that it plans to report on in the Fall of 2009.

that dealerships were ordering fewer new vehicles, which caused a cash-flow problem for the OEMs, and consumers deferred new vehicle purchases, which caused further problems for both dealers and OEMs, not only in the United States but essentially globally as well. The dire situation of Chrysler and General Motors in late 2008 and early 2009, including their recent bankruptcies and inability to secure credit for continuing operations, led to the federal government taking a role by investing in the industry.

Simultaneously with the economic problems, the auto industry is also experiencing the effects of new regulations related to fuel economy and the environment. New corporate average fuel economy (CAFE) standards enacted in late 2007 set an aggressive time line for improvement in vehicle fuel economy regardless of oil prices and resulting market demand. At the same time, the regulation of carbon dioxide (CO_2) emissions that was being enacted, along with actions in some states, could have necessitated either using lower-carbon fuels or achieving higher fuel economy levels to meet requirements for CO_2 reductions. Recent rules proposed by the Administration, apparently developed with participation by the OEMs, will result in even more aggressive fuel mileage requirements, which may become uniform nationwide. In addition to the major changes noted above, two already existing trends in the auto industry include a shift to decentralization of operations and internationalization of the industry. Decentralization has meant increasing automotive supplier involvement in contributing to the R&D for product development. Further, many supplier activities that were owned and operated by individual auto companies were spun off into independent companies, again decreasing automakers' internal R&D activities. One result has been a supplier base that moves more toward the center of innovation. Further, common components have become available for the industry at large, thus opening the door to commodity components and systems.

Additionally, foreign auto companies have taken an increasing share of the U.S. auto market, many benefiting from government-sponsored research in their home markets. At the same time the U.S. auto industry is expanding operations outside the United States, which increases the potential for leakage of U.S.-developed technology into emerging markets.

A final consideration is that current fiscal realities raise the unsettling possibility that U.S. industry might, at least in the near term, be unable to continue developing long-term solutions like hydrogen/fuel cell vehicles. The accumulated experience and expertise could be lost as researchers disperse, leaving the Partnership with the more challenging task of developing fuel-cell and other long-term advanced technology options without automobile industry collaboration. Conversely, a robust Partnership with full participation by the domestic auto industry has the potential to be an important factor in a renaissance of the U.S. auto industry.

SUGGESTED ACTIONS

The Partnership should consider adapting its goals and strategy in response to changing U.S. priorities and new findings (e.g., NRC. 2008b). An increased emphasis is suggested by the committee on the R&D needed to produce usable short-term technologies (e.g., better batteries for PHEVs, improved ICEs), along with continuing R&D on the long-term technologies (e.g., BEVs, cellulosic ethanol and other non-food- crop biofuels, hydrogen fuel, and fuel cells). An increased emphasis is also required for technologies that will produce significantly lower greenhouse gas emissions (e.g., CO_2) and the increased use of domestic energy sources, especially biofuels. As noted previously, some of the goals of the President's *New Energy for America* plan are to save significant amounts of oil within 10 years, put 1 million plug-in hybrid vehicles on the road by 2015, reduce greenhouse gas emissions by 80 percent by 2050, and make significant investments in climate-friendly energy development and deployment over the next 10 years (Satyapal and Davis, 2009a).

Goals and Strategy

Overall the strategy of the Partnership (see Attachment IV) seems to be appropriate and should generally continue, but with some modifications. Specifically, the committee considers that the government-industry partnership is working well and should continue (NRC, 2000, 2001, 2005, 2008a). However, the Partnership should consider whether the timeline for the long-term goals (hydrogen infrastructure/fuel cells) should be extended and more emphasis placed on nearer-term technologies. The latter can possibly help revive the industry and also help with societal issues, such as environmental concerns associated with greenhouse gas emissions, economic concerns associated with massive imports of crude oil, and social issues associated with the loss of many thousands of auto industry jobs.

In addition, the spectrum of needed technologies and the range of applicable time scales suggest the utility of developing new models for stimulating private sector researchers, national laboratory scientists, and academics to engage in new productive collaborations. It is also crucial to attract good students to these research efforts, partly to enhance and restore U.S. scientific leadership, but also to entice the best and brightest of new generations to contribute to long-term energy and environmental solutions.

These factors suggest that support for U.S. industry such as that provided by the Partnership is probably needed now more than ever. With the Administration's goals in mind and given uncertainties about the cost, performance, and consumer acceptance of many of the vehicle technologies under development, it is vital for the Partnership to have a diverse portfolio of options. As is noted in Conclusion 1 of a recent NRC (2008b) report, "A portfolio of technologies including hydrogen fuel cell vehicles, improved efficiency of conventional vehicles, hybrids, and the use of biofuels—in conjunction with required new public drivers—has the potential to nearly eliminate gasoline use in light-duty vehicles by the middle of the century, while reducing fleet greenhouse gases to less than 20 percent of their current levels" (p. 4). The Partnership should not lose focus on its main goals of providing good management and oversight of all its activities. However, depending on congressional actions and Administration directives, as well as budgets and funding, the Partnership should consider the following:

- Rewording its mission statement and goals to reflect consistency with the new Administration's goals and priorities.
- Not abandoning programs on the long-term high-risk vehicle technologies that could be highly transformational with regard to reduced use of petroleum and reduced emissions, namely, fuel cells, hydrogen storage, and batteries for BEVs, as well the exploration of innovative systems concepts.
- Incorporating the broader scope of a "cradle-to-grave" analysis rather than a "well-to-wheels" approach in program planning from production to recycle to better consider total energy consumption, total emissions, and total environmental impact.[6]
- Emphasizing R&D to support development of nearer-term technologies (such as advanced ICEs, and better batteries for HEVs and PHEVs) and long-term technologies (such as cellulosic-based and other non-food-crop biofuels/ICEs, hydrogen/fuel cell vehicles, and all-electric vehicles) and define a transition pathway from nearer-term to long-term technologies, including targets, milestones, and go/no-go decision points.
- Expanding efforts to support exploratory projects on transformative and revolutionary ideas that are beyond the current scope of the Partnership.[7] This should include, if possible, joint funding and cooperation with different DOE offices, and the enlistment of a broad group of stakeholders

---

[6] "Well-to-wheels" in the context of motor vehicles commonly refers to an analysis covering fuel production to fuel usage in the vehicle. For example, in Argonne National Laboratory's GREET model, such analyses have been conducted for biofuels, electricity, and gasoline. Life-cycle analyses (LCA) would include the production of vehicles, including energy storage technologies (e.g., batteries), and the distribution of vehicles as well as the disposal of components after the useful life of the vehicle has been reached, which are not included in the "well-to-wheels" analysis. However, a comprehensive "cradle-to-grave" analysis would include all these aspects. Unlike gasoline, some fuels do not come from wells, and hence some prefer the term "source-to-wheels." Similarly, those interested in complete recycle (which is probably not practically possible) use "cradle to cradle" instead of "cradle to grave."

[7] There may be opportunities to leverage the newly formed Advanced Research Projects Agency–Energy (ARPA-E), a new DOE organization created specifically to foster R&D of transformational energy-related technologies. See http://www.energy.gov/news2009/documents2009/ARPA-E_FOA.pdf.

including academia, start-ups,[8] and mature companies and providing them with support for at least minimal R&D efforts.

- Maintaining stable funding because of its importance for fuel production and delivery activities. It is the committee's view that it is a critical to understand and address the barriers, costs, and environmental impacts not only of hydrogen but also of other potential energy carriers and fuels as well.
- Finally, the Partnership should consider broadening the scope of the technical approaches being considered within each of the three major fuel and vehicle technology pathways (biofuels/ICEs, PHEVs/BEVs, and hydrogen/fuel cell vehicles). In the electric vehicle area, other storage approaches such as nano-enhanced capacitors and batteries beyond those with lithium chemistries should be the subject of basic and potential future applied research. In addition, many fuel cell approaches and hydrogen storage options should continue to be investigated, and options should not be prematurely shut down.

The committee recognizes that many of the actions it might see as desirable for pursuing revised goals involve primarily the DOE but cannot be implemented unilaterally by the DOE. It also recognizes that successful R&D alone does not necessarily translate into the commercialization of advanced vehicles and fuels and entry into the marketplace. Further, even with commercialization, substantial penetration into the general U.S. economy is required if significant reductions are to be achieved in petroleum consumption and greenhouse gas emissions. Achieving such goals will require an enthusiastic "buy-in" by the private sector.

Many components of the private sector, in addition to OEMs and major suppliers, can make meaningful contributions to technology advancements. In the committee's view, the innovation capacity of the private sector is best motivated through consistent and predictable policy and market incentives. For example, innovators, entrepreneurs, and investors rely on stable policies and incentives to evaluate the risks and benefits of pursuing alternative technologies and thereby allocate private resources efficiently.

An environment of stable and predictable incentives for vehicle technologies could be created in several ways, including the establishment of predictable carbon prices, a carbon trading plan, performance standards, and policies or incentives to reduce energy imports, as well as commitments with long enough timescales to encourage active participation.[9] The committee recognizes that the Partnership does not have complete control of the broader market and policy environment. It

[8] Note that start-up companies typically need technology maturation funds (i.e., proof-of-concept and product-development support) more than funds for R&D.

[9] For a recent discussion of the policy issues related to the market adoption of high-risk technologies like fuel cell vehicles and the associated hydrogen fuel infrastructure, see NRC (2008b).

simply notes that such policy signals can provide important "market pull" to aid in the deployment of technologies arising from the "technology push" of R&D programs. Stable incentives are also pointed out because of their importance, in the committee's view, in enabling the nation to realize the rapid transformation in vehicle technologies that appears to be an Administration priority.

National policy clearly seeks that the manufacturing of advanced new vehicles occur in the United States to the greatest extent possible. But for this goal to be realized, U.S. manufacturing will need a durable, structural advantage to compete effectively in world markets.

This competitive advantage could derive from two sources: (1) design of vehicle systems and components to improve manufacturability; and (2) general research in manufacturing technologies and processes to develop competitive manufacturing advantages. High-priority areas for vehicle systems and components in which proprietary competitive advantage could be gained, which the committee suggests be actively pursued, include both component technology advancements and advanced manufacturing processes for:

- Lithium-ion and other promising batteries,
- Power electronics, including packaging, and
- Advanced ICEs.

In addition the committee also suggests continuing similar efforts on long-term technologies including:

- Fuel cell and stack components, and
- Vehicle onboard hydrogen storage.

The committee also suggests that the lightweight materials program continue to be an integral part of the efforts for improving fuel efficiency for vehicles that would utilize both nearer-term and longer-term technologies.

Implementation

The following are some committee suggestions for the Partnership for implementing efforts to deal with the changes in the automotive sector and the goals the Administration is pursuing:

- Consider temporary reductions in cost-share requirements for a number of joint program efforts. Many universities and small industrial organizations could have considerable difficulty, under current economic conditions, in providing matching funds as currently required.
- Consider directly funding the OEMs to keep their in-house, non-petroleum research, development and demonstration (RD&D) pro-

grams active (e.g., RD&D on hydrogen/fuel cells, BEVs, PHEVs, and biofueled ICEs).

- As the Partnership moves toward possible commercialization decisions for a technology, consider allocating more funding to private sector companies to seed development of a robust, strong supply base in these developing advanced technologies.
- Consider whether the Partnership might benefit from exploring systematic linkages with entrepreneurs and innovators whose contributions could accelerate the pace of innovation in the industry. Equally important, such new ventures might benefit from systematic contact with the markets provided by current Partnership members.
- Assess Partnership member makeup and collaborations. New members added to the Partnership can bring expertise in important areas and contribute to reaching goals. There is also the danger that too many non-contributing members can lead to a more cumbersome and inefficient operation that is more difficult to manage. If new members and/or other active participants are added, the Partnership should consider including:

    1. Electric Power Research Institute (EPRI) representative(s) and DOE representatives that are involved with Smart Grid activities, for the utilities technical team;
    2. Biofuels representative(s), for the production and delivery technical teams;
    3. More involvement and coordination with appropriate representatives from the U.S. Department of Transportation, especially in areas that will interact with Partnership long-term goals; and
    4. Expertise in and activities relevant to manufacturing processes, including relevant supplier industries, among the technical programs.

- Expand support for next-generation research being performed at current and future automotive suppliers as well as OEMs.

Once again, the committee appreciates the opportunity to have provided some suggestions to the Partnership as it moves forward in these challenging times.

Sincerely,

Vernon P. Roan, *Chair*
Committee on Review of the Research Program of the
FreedomCAR and Fuel Partnership, Phase 3

Cc: Patrick Davis, DOE Office of Vehicle Technologies
Keith Hardy, DOE Office of Vehicle Technologies
Carl Maronde, DOE National Energy Technology Laboratory
Ken Howden, DOE Office of Vehicle Technologies

Attachments

I. Committee on Review of the Research Program of the FreedomCAR and Fuel Partnership, Phase 3
II. Statement of Task
III. Presentations and Discussions with Representatives of the Partnership at the Committee Meeting, April 27, 2009
IV. FreedomCAR and Fuel Partnership

REFERENCES

Chu, S. 2009. U.S. Department of Energy FY2010 Budget: Transforming the Energy Economy Through Science and Innovation (May 7). Available at http://www.energy.gov/media/Secretary_Chu_2010_Budget_rollout_presentation.pdf.

DOE (U.S. Department of Energy). 2009a. *Addendum to the FreedomCAR and Fuel Partnership Plan to Integrate Electric Utility Industry Representatives*. February. Washington, D.C.: U.S. Department of Energy, Energy Efficiency and Renewable Energy. Available at http://www1.eere.energy.gov/vehiclesandfuels/pdfs/program/fc_fuel_addendum_2-09.pdf.

DOE. 2009b. U.S. Department of Energy's FY2010 Budget Request to Congress. Available at http://www.cfo.doe.gov/crorg/cf30.htm.

DOE. 2006. *Partnership Plan: FreedomCAR & Fuel Partnership*. March. Washington, D.C.: U.S. Department of Energy, Energy Efficiency and Renewable Energy. Available at http://www.eere.energy.gov/vehiclesandfuels/pdfs/program/fc_fuel_partnership_plan.pdf

DOE (U.S. Department of Energy). 2004a. *Hydrogen, Fuel Cells and Infrastructure: Multi-Year Research, Development and Demonstration Plan*. DOE/GO-102003-1741. Washington, D.C.: U.S. Department of Energy, Energy Efficiency and Renewable Energy. Available at http://www.eere.energy.gov/hydrogenandfuelcells/mypp/.

DOE. 2004b. *FreedomCAR and Vehicle Technologies Multi-Year Program Plan*. Washington, D.C.: U.S. Department of Energy, Energy Efficiency and Renewable Energy. Available from EERE Information Center at www.eere.energy.gov.

NRC (National Research Council). 2008a. Review of the Research Program of the FreedomCAR and Fuel Partnership, Second Report. Washington, D.C.: The National Academies Press.

NRC. 2008b. Transitions to Alternative Transportation Technologies—A Focus on Hydrogen. Washington, D.C.: The National Academies Press.

NRC. 2005. Review of the Research Program of the FreedomCAR and Fuel Partnership, First Report. Washington, D.C.: The National Academies Press.

NRC. 2001. *Review of the Research Program of the Partnership for a New Generation of Vehicles, Seventh Report*. Washington, D.C.: National Academy Press.

NRC. 2000. *Review of the Research Program of the Partnership for a New Generation of Vehicles, Sixth Report*. Washington, D.C.: National Academy Press.

NRC. 1998. Effectivenes of the United States Advanced Battery Consortium as a Government-Industry Partnership. Washington, D.C.: National Academy Press.

PNGV (Partnership for a New Generation of Vehicles). 1995. *Partnership for a New Generation of Vehicles Program Plan* (draft). Washington, D.C.: U.S. Department of Commerce, PNGV Secretariat.

Satyapal, S., and P. Davis. 2009a. Changes in Objectives of the Partnership. Presentation to the Committee on Review of the Research Program of the FreedomCAR and Fuel Partnership, Phase 3, April 27, Washington, D.C.

Satyapal, S., and P. Davis. 2009b. Budget Outlook (FY09-10). Presentation to the Committee on Review of the Research Program of the FreedomCAR and Fuel Partnership, Phase 3, April 27, Washington, D.C.

The White House. 1993. *Historic Partnership Forged with Automakers Aims for Threefold Increase in Fuel Efficiency in as Soon as Ten Years.* Washington, D.C.: The White House.

ACKNOWLEDGMENT OF REVIEWERS

This letter report was reviewed in draft form by the following individuals, chosen for their diverse perspectives and technical expertise in accordance with procedures approved by the National Research Council's Report Review Committee: Allen J. Bard (NAS), University of Texas, Austin; John Heywood (NAE), Massachusetts Institute of Technology; Thomas M. Jahns, University of Wisconsin, Madison; Trevor Jones (NAE), ElectroSonics Medical, Inc.; William F. Powers (NAE), consultant, Ford Motor Company (retired); Robert W. Shaw, Jr., Aretê Corporation; Dan Sperling, University of California, Davis; and Jay Whitacre, Carnegie Mellon University. The review was overseen by Granger Morgan (NAS), Carnegie Mellon University. Although the individuals listed above provided many constructive comments and suggestions, they were not asked to endorse the report's conclusions or recommendations, nor did they see the final draft of the report before its release. Responsibility for the final content of this report rests entirely with the authoring committee and the institution.

ATTACHMENT I

COMMITTEE ON REVIEW OF THE RESEARCH PROGRAM OF THE FREEDOMCAR AND FUEL PARTNERSHIP, PHASE 3

VERNON ROAN, *Chair,* University of Florida, Director, Center for Advanced Studies in Engineering and Professor of Mechanical Engineering (retired), Gainesville

DEBORAH LYNN BLEVISS, Independent Consultant, Falls Church, Virginia

DAVID BODDE, Senior Fellow and Professor, Clemson University, South Carolina

KATHRYN BULLOCK, President and Owner, Coolohm, Inc., Blue Bell, Pennsylvania

HARRY COOK, NAE,[1] Professor Emeritus, University of Illinois (retired)

GLENN EISMAN, Principal Partner, Eisman Technology Consultants, LLC, Niskayuna, New York
W. ROBERT EPPERLY, Consultant, Mountain View, California
WILLIAM ERNST, Owner, EnerSysCon, Troy, New York
DAVID FOSTER, Professor of Mechanical Engineering, University of Wisconsin-Madison, Engine Research Center
GERALD GABRIELSE, NAS,[2] Leverett Professor of Physics, Harvard University
LINOS JACOVIDES, Director, Delphi Research Labs (retired), Grosse Point Farms, Michigan
HAROLD H.C. KUNG, Professor of Chemical Engineering, and Director, Northwestern University Center for Catalysis and Surface Science, Evanston, Illinois
CHRISTOPHER MAGEE, NAE, Professor, Engineering Systems Division, Massachusetts Institute of Technology, Cambridge
CRAIG MARKS, NAE, Vice President, Technology and Productivity, AlliedSignal, Inc. (retired), Bloomfield Hills, Michigan
GENE NEMANICH, Vice President, Chevron Hydrogen Systems (retired)
BERNARD ROBERTSON, NAE, Senior Vice President, Engineering Technologies and Regulatory Affairs; and General Manager-Truck Operations, Chrysler LLC (retired), Bloomfield Hills, Michigan
R. RHOADS STEPHENSON, Caltech/Jet Propulsion Laboratory (retired), La Cañada, California
KATHLEEN TAYLOR, NAE, Director, Materials and Processes Laboratory, General Motors Corporation (retired), Fort Myers, Florida
BRIJESH VYAS, Distinguished Member of Technical Staff, LGS Innovations, Warren, New Jersey

National Research Council Staff

JAMES ZUCCHETTO, Director, Board on Energy and Environmental Systems (BEES)
LANITA JONES, Administrative Coordinator, BEES
DANA CAINES, Financial Associate, BEES

[1]NAE=member, National Academy of Engineering.
[2]NAS=member, National Academy of Sciences.

ATTACHMENT II

STATEMENT OF TASK

The National Academies' National Research Council (NRC) Committee on Review of the Research Program of the FreedomCAR and Fuel Partnership, Phase 3, will address the following tasks **(Note: the interim letter report will address Task 6)**:

(1) Review the challenging high-level technical goals and timetables for government and industry R&D efforts, which address such areas as (a) integrated systems analysis; (b) fuel cell power systems; (c) hydrogen storage systems; (d) hydrogen production and distribution technologies necessary for the viability of hydrogen-fueled vehicles; (e) the technical basis for codes and standards; (f) electric propulsion systems; (g) electric energy storage technologies; (h) lightweight materials; and (i) advanced combustion and emission control systems for internal combustion engines (ICEs).
(2) Review and evaluate progress and program directions since the Phase 1 and 2 reviews toward meeting the Partnership's technical goals, and examine ongoing research activities and their relevance to meeting the goals of the Partnership.
(3) Examine and comment on the overall balance and adequacy of the research and development effort, and the rate of progress, in light of the technical objectives and schedules for each of the major technology areas.
(4) Examine and comment, as necessary, on the appropriate role for federal involvement in the various technical areas under development, especially in light of activities ongoing in the private sector or in the states.
(5) Examine and comment on the Partnership's strategy for accomplishing its goals, especially in the context of ongoing developments in biofuels, plug-in hybrid electric vehicles, electric vehicles, the recent enactment of legislation on corporate average fuel economy standards for light-duty vehicles, and possible legislation on carbon emissions. Other issues that the committee might address include (a) program management and organization; (b) the process for setting milestones, research directions, and making Go/No Go decisions; (c) collaborative activities needed to meet the program's goals (e.g., among the various offices and programs in DOE, the U.S. Department of Transportation, USCAR, the fuels industry, electric power sector, universities, other parts of the private sector [such as venture capitalists], and others); and (d) other topics that the committee finds important to comment on related to the success of the program in meeting its technical goals.
(6) As a first step in examining the Partnership's strategy, and given the changes that may take place with the new Administration, the committee at its first full committee meeting will address potential changes in the program strategy and program structure. The committee will write a short interim letter report with suggestions and recommendations on program strategy

and structure and aim to deliver it to the sponsor within 1 month after the meeting. The date of delivery of the letter report will be contingent on when the meeting is scheduled and timely input of information from the representatives of the Partnership.

(7) Review and assess the actions that have been taken in response to recommendations from the NRC Phase 2 review of the Partnership.

(8) Write a final report documenting its conclusions and recommendations.

ATTACHMENT III

PRESENTATIONS AND DISCUSSIONS WITH REPRESENTATIVES OF THE PARTNERSHIP AT THE COMMITTEE MEETING, APRIL 27, 2009

Changes in Objectives in the Partnership
Sunita Satyapal and Patrick Davis, Office of Energy Efficiency and Renewable Energy (EERE), U.S. Department of Energy (DOE)

- General priorities being set by the Secretary
- Major initiatives that might influence Partnership

Budget Outlook (FY09-10)
Sunita Satyapal and Patrick Davis, EERE, DOE

- Overall Level of Resources
- How are Resource Allocations Changing?
- How Does Stimulus Plan/Loan Program Affect Partnership?

Overview and Progress & Outlook on Existing Program Efforts
Sunita Satyapal and Patrick Davis, EERE, DOE

- Progress and outlook for meeting future targets, especially with regard to critical technologies, and optimistic and pessimistic views and the need for (or lack of) changes in projected timescales
- Where has the program made significant progress?
- Where has progress not been adequate?
- Do targets and milestones need to be changed?

Automotive Industry Partners' Views on Progress, Strategy, Future Outlook, and Structure of Program
William Peirce, General Motors Corporation; Reginald Modlin, Chrylser LLC; and John Sakioka, Ford Motor Company

Fuel Industry Partners' Views on Progress, Strategy, Future Outlook, and Structure of Program
George Parks, ConocoPhillips; Puneet Verma, Chevron Technology Ventures; and James Kegerreis, ExxonMobil

Utility Industry Partners' Views on Progress, Strategy, Future Outlook, and Structure of Program
Robert Graham, Southern California Edison

DOE's Views on Progress, Strategy, Future Outlook, and Structure of Program
Sunita Satyapal and Patrick Davis, EERE, DOE

ATTACHMENT IV

FREEDOMCAR AND FUEL PARTNERSHIP

History and Background

The FreedomCAR and Fuel Partnership is a research and development (R&D) program designed to enable long-range, significant changes in automobiles and their energy supply systems for the purpose of obtaining major societal benefits, such as reduced petroleum consumption and reduced levels of harmful gaseous emissions to the atmosphere. Research projects sponsored at government laboratories, universities, and private companies are chosen and monitored by joint industry/government technical teams. This structure helps focus expenditures on research to support projects that are relevant to the long-range, pre-competitive research needs envisioned by automotive, energy, and, now, utility companies, and help to meet the nation's societal needs as articulated by the government. The basic structure has evolved and has improved over almost 15 years and has proven to be an excellent mechanism for achieving progress (NRC, 2000, 2001, 2005, 2008a).

The DOE has been involved for about 30 years in R&D programs related to advanced vehicular technologies and alternative transportation fuels. During the 1990s, much of this R&D was conducted as part of the Partnership for a New Generation of Vehicles (PNGV) program, which was formed between the federal government and the auto industry's USCAR.[10] Building on the PNGV program,

[10] USCAR, which predated the formation of PNGV, was established by Chrysler Corporation, Ford Motor Company, and General Motors Corporation. Its purpose was to support intercompany, precompetitive cooperation that would reduce the cost of redundant R&D, especially in areas mandated by government regulation, and make the U.S. industry more competitive with international companies. Chrysler Corporation merged with Daimler Benz in 1998 to form DaimlerChrysler. In 2007, DaimlerChrysler divested from a major interest in the Chrysler Group and Chrysler LLC was formed; DaimlerChrysler was renamed Daimler AG. The PNGV sought to significantly improve the

in January 2002 the Secretary of Energy and executives of DaimlerChrysler, Ford, and General Motors announced a new government-industry partnership between DOE and USCAR called FreedomCAR, with CAR standing for Cooperative Automotive Research. In September 2003, FreedomCAR was expanded to include the five large energy companies mentioned previously to address issues related to the supporting fuel infrastructure. The expanded partnership is called the FreedomCAR and Fuel Partnership (DOE, 2006).[11] During the time period since the last NRC phase 2 review (NRC, 2008a), the Partnership has expanded to include the utility industry (as noted previously, DTE Energy and Southern California Edison) (DOE, 2009a). These new partners were added to address issues associated with use of the electric transmission and distribution systems that would accompany commercial deployment of PHEVs and BEVs.

The Partnership addresses the development of advanced technologies for all light-duty passenger vehicles. It also addresses technologies for hydrogen production, distribution, dispensing, and storage. Funding for research, development, and demonstration activities goes to the national laboratories, private companies, and universities. Especially in the case of development activities, projects costs are often shared between the private sector and the federal government.

The Partnership plays an important role in the planning, pursuit, and assessment of high-risk R&D for many of the needed vehicle and fuel technologies, and federal funds allow much of this work to move forward. It also serves as a communication mechanism for the interested players, including government, the national laboratories, private industry, universities, the public, and others. This structure recognizes both the long-range, high-risk research needs envisioned by automotive and energy companies, and the nation's societal needs related to automotive vehicles and fuels, as articulated by government, in defining the appropriate goals and selecting the best way of achieving them. This capability is seen by the committee as a major strength of the Partnership that should be retained even if other changes are made.

---

nation's competitiveness in the manufacture of future generations of vehicles, to implement commercially viable innovations emanating from ongoing research on conventional vehicles, and to develop vehicles that achieve up to three times the fuel efficiency of comparable 1994 family sedans (NRC, 2001; PNGV, 1995; The White House, 1993).

[11] In February 2003, before the announcement of the FreedomCAR and Fuel Partnership, the President announced the FreedomCAR and Hydrogen Fuel Initiative to develop technologies for (1) fuel efficient motor vehicles and light trucks, (2) cleaner fuels, (3) improved energy efficiency, and (4) the hydrogen production and nationwide distribution infrastructure needed for vehicle and stationary power plants, to fuel both hydrogen ICEs and fuel cells (DOE, 2004a). The expansion of the FreedomCAR and Fuel Partnership to include the energy sector after the announcement of the initiative also supports the goal of the Hydrogen Fuel Initiative.

Current Structure of the Partnership

The administrative structure of the Partnership includes the Executive Steering Group, which oversees the Joint Operations Group, Fuels Operations Group, FreedomCAR Operations Group, and the newly added Utility Operations Group. The DOE managers and respective energy company, automotive companies (OEMs), and utility directors of these groups oversee technical teams that are responsible for developing R&D plans and roadmaps, reviewing research results, and evaluating technical progress toward meeting research goals. Realizing that there will be a portfolio of energy carriers (fuels) and mobility technologies necessary to move forward, and to address the technical challenges associated with the different fuel/vehicle technology pathways, the Partnership has established a technical roadmap with specific, quantitative 2010 and 2015 technology and cost goals in eight areas:

- Internal combustion engines (both petroleum and hydrogen fueled),
- Fuel cell power systems,
- Fuel cells,
- Hydrogen storage systems,
- Energy storage systems for hybrid vehicles,
- Hydrogen production and delivery systems,
- Electric propulsion systems, and
- Materials for lightweight vehicles.

It is within this structure that the Partnership sets priorities, determines technical targets and milestones, and performs the research attempting to achieve those targets. Regular reviews, both internal and external, are conducted to receive feedback and critiques of individual and group projects.

Appendix C

Organizational Chart

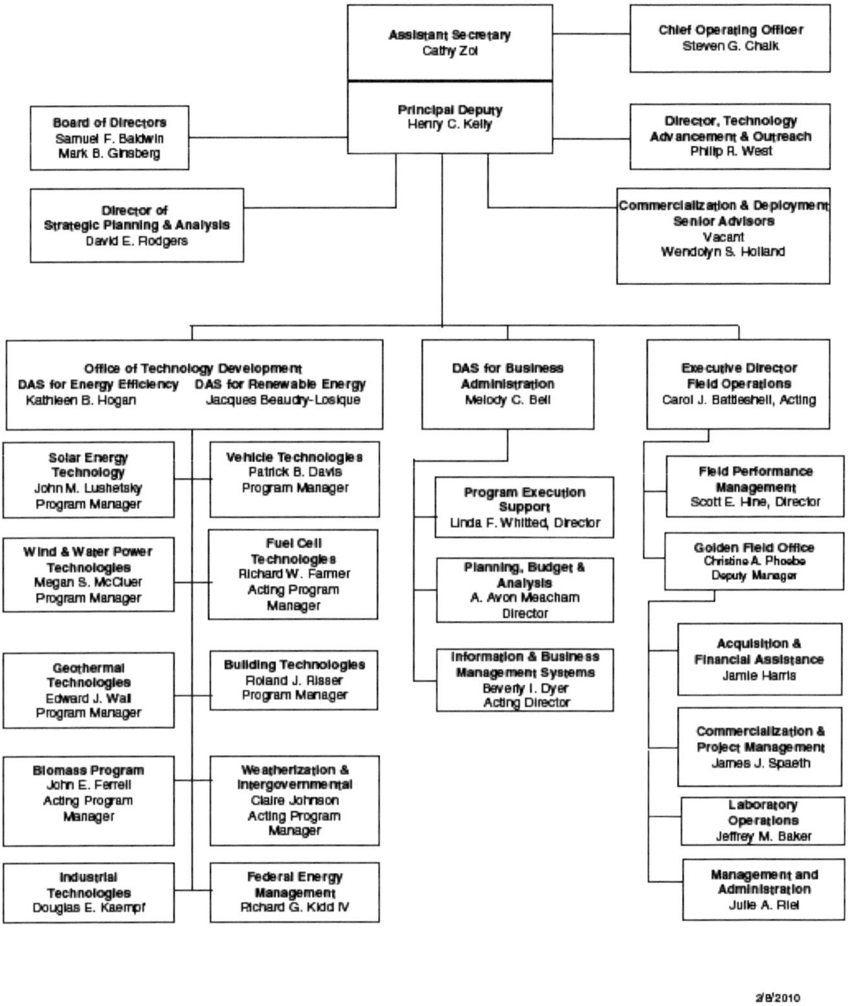

FIGURE C-1 Organizational chart for the U.S. Department of Energy's Office of Energy Efficiency and Renewable Energy (as of February 8, 2010).

Appendix D

Recommendations from National Research Council Review of the FreedomCAR and Fuel Research Program, Phase 2

CHAPTER 2: MAJOR CROSSCUTTING ISSUES

Strategic Planning and Decision Making

Recommendation [2-1]. DOE should accelerate the development and validation of modeling tools that can be used to assess the roles of various propulsion system and vehicle technologies and fuels, and utilize them to determine the impact of the various opportunities on the overall Partnership goals of reducing petroleum use and air pollutant and greenhouse gas emissions. Sensitivity analysis, from worst case to optimistic scenarios, should be performed to assess these impacts. Models, input data, and assumptions should be independently reviewed in order to validate and refine the models and lend credibility to the conclusions derived from them.

Recommendation [2-2]. The FreedomCAR and Fuel Partnership should use its technical and cost systems analysis capabilities as an essential component in program management to assess progress in meeting technical and cost targets, to examine the impact of not meeting those targets, to adjust program priorities, and to make go/no-go decisions.

Recommendation [2-3]. The FreedomCAR and Fuel Partnership's Executive Steering Group should establish a high-level planning group to develop a strategic plan appropriate for the next phase of the nation's collaborative vehicle and fuels technology R&D program.

Recommendation [2-4]. The Partnership management should assess how best to pursue PHEV technology within the FreedomCAR and Fuel Partnership program and determine the cost and performance merits relative to hydrogen fuel cell vehicles using the same vehicle structural weight for both systems.

Recommendation [2-5]. DOE should utilize its modeling capability to assess the impact of market interventions on both the technical goals of the FreedomCAR and Fuel Partnership, and their overall potential impact, and use these assessments to inform the R&D planning process.

Recommendation [2-6]. The Partnership should evaluate the potential for analyzing and predicting market responses to the vehicle technologies and fuels that may result from Partnership efforts to better inform its assessments of the new technologies that are likely to be needed to meet the nation's goal of reducing petroleum consumption and greenhouse gases.

Safety

Recommendation [2-7]. DOE should establish a program to address all end-to-end safety aspects as well as codes and standards. Such a program could be viewed as an extension of the current quantitative risk analysis activity, which is focused on the filling station. This task should be adequately funded and expanded. The priority for expansion should go to (1) the vehicle and (2) the fuel distribution system.

Recommendation [2-8]. The Department of Transportation (DOT), including all relevant entities within DOT, should develop a long-range, comprehensive hydrogen safety plan with budget estimates and milestones to 2015. The milestones developed in this plan should be integrated into the codes and standards technical team milestones and roadmap.

Recommendation [2-9]. The codes and standards technical team should extend the planning horizon in its plan to 2015, integrate the DOT milestones into its own milestones and roadmap, and make the safety and codes and standards milestones consistent with funding levels and progress to date.

Recommendation [2-10]. DOE should establish a program to collect and analyze failure data and field experience including data from the National Highway Traffic Safety Administration on compressed natural gas (CNG) and hydrogen components, subsystems, vehicles, and fueling stations.

Recommendation [2-11]. DOE should convene a review by a panel of independent outside experts of the hydrogen compatibility of materials, prioritize the materials to be tested, taking into account the likelihood of their application, and review test procedures and conditions.

Recommendation [2-12]. DOE should accelerate work on delayed ignition of unintended hydrogen releases, including in parking structures and tunnels, in support of various efforts to develop and revise building codes.

Technical Validation

Recommendation [2-13]. DOE should continue to disseminate the results of the technical validation activity to supporting organizations outside the Partnership in order to promote widespread innovation and competition. DOE management needs to systematically evaluate the information being generated by each project to determine when the project should be terminated based on its relevance and on the value of the information. On the other hand, DOE management should not prematurely drop support for the overall technical validation and learning demonstrations as their importance cannot be overemphasized. DOE and the Partnership should develop a long-range plan for technology validation that continues to at least 2015.

Recommendation [2-14]. DOE management should maintain adequate support for technical validation as it is essential to the overall Partnership. This support should be balanced and cover both the vehicles themselves and the fuel infrastructure needed. To achieve the rapid learning that the overall project requires, DOE should also keep the validation activities focused on their primary purpose—the accumulation, analysis, and dissemination of experience from the field. Safety should be stressed throughout the learning demonstration program, because an accident early on could attract publicity out of proportion to its true consequences.

Building Partnerships with New Ventures

Recommendation [2-15]. DOE should conduct a systematic assessment of the success (or failure) of all its SBIR/STTR-funded companies rather than selected case studies.

Recommendation [2-16]. The Partnership should seek ways beyond the SBIR and STTR programs to improve communications between it and the entrepreneurial community.

CHAPTER 3: VEHICLE SUBSYSTEMS

Advanced Combustion, Emissions Control, and Hydrocarbon Fuels

Recommendation [3-1]. The Partnership should formulate and implement a clear set of criteria to identify and provide support to ICE combustion and emission control projects that are precompetitive and show potential for improvements well beyond those currently being developed by industry.

Recommendation [3-2]. DOE should actively encourage collaborations among the national laboratories, industry, and academia to more effectively direct research efforts to areas where enhanced fundamental understanding is most needed.

Recommendation [3-3]. The Partnership should investigate the impact on emissions of combustion mode switching and transient operation with LTC.

Recommendation [3-4]. The Partnership should perform a detailed analysis of the potential improvement in efficiency and the cost-effectiveness of the exhaust gas heat recovery effort and make a go/no-go decision about this work.

Fuel Cells

Recommendation [3-5]. The Partnership should conduct sensitivity analyses on key fuel cell targets to determine the trade-offs and tolerances in engineering specifications allowable while still meeting fuel cell vehicle engineering requirements.

Recommendation [3-6]. The Partnership should reassess the current allocation of funding within the fuel cell program and reallocate it as appropriate, in order to prioritize and emphasize the R&D that addresses the most critical barriers. In particular, the Partnership should give membranes, catalysts, electrodes, and modes of operation the highest priority. In particular, it should also

- Place greater emphasis on science and engineering at the cell level and, from a systems perspective, on integration and subcomponent interactions;
- Reduce research on carbon-based supported catalysts in favor of developing carbon-free electrocatalysts;
- Ensure that BES funding of membranes, catalysts, and electrodes remains a high priority of the program; and
- Apply go/no-go decision making to stationary fuel cell system initiatives that are not directly related to transportation technologies.

Onboard Hydrogen Storage

Recommendation [3-7]. The hydrogen storage program should continue to be supported by the Partnership at a high level since finding a suitable storage material is critical to fulfillment of the vision for the hydrogen economy. Both basic and applied research should be conducted.

Recommendation [3-8]. The Partnership should rebalance the R&D program for hydrogen storage to shift resources to the more promising approaches as knowledge is gained. The new systems engineering center of excellence should look at all of the system requirements simultaneously, not just the system weight percent storage goal, and guide this rebalancing.

Recommendation [3-9]. In the event that no onboard hydrogen systems are found that are projected to meet targets, the Partnership should perform appropriate studies to determine the risks and consequences of relying on pressurized hydrogen storage. They should include production and delivery issues as well as effects on vehicle performance, safety, and costs.

Recommendation [3-10]. The Partnership should pursue research leading to lower costs for high-quality carbon fibers and bonding materials that would allow higher operating temperatures for compressed hydrogen gas storage.

Recommendation [3-11]. The Partnership should maintain a strong basic research activity on hydrogen storage. New hydrogen storage concepts should continue to be supported by the Office of Basic Energy Sciences.

Electrochemical Energy Storage

Recommendation [3-12]. The Partnership should conduct a thorough analysis of the cost of the Li-ion battery for each application; hybrid electric vehicles (HEVs), PHEVs, battery electric vehicles (EVs), and hydrogen-fueled fuel cell HEVs. The analysis should re-examine the initial assumptions, including those for both market forces and technical issues, and refine them based on recent materials and process costs. It should also determine the effect of increasing production rates for the different systems under development.

Recommendation [3-13]. The Partnership should significantly intensify its efforts to develop high-energy batteries, particularly newer, higher specific energy electrochemical systems within the long-term battery research subactivity and in close coordination with BES. High-energy batteries provide the surest way to successful batteries for PHEVs.

Recommendation [3-14]. The Partnership should move forward aggressively with completing and executing its R&D plan for plug-in hybrid electric vehicles.

Electric Propulsion, Electrical Systems, and Power Electronics

Recommendation [3-15]. The Partnership should conduct a meta-analysis and develop quantitative models to identify fundamental geometric limitations that ultimately set bounds on and lead to the realization of the size, mass, and cost of power converters and electric propulsion systems in relation to the physical properties of materials and processes such as dielectric strength, magnetic saturation, thermal conductivity, etc. This will allow the various ongoing and future efforts to be benchmarked against the theoretical boundaries of what is possible and enable the establishment of appropriate directions in research goals.

Recommendation [3-16]. In general, the Partnership should focus on the projects that address specific performance and cost goals of the program on the basis of the results of the meta-analysis recommended above. Specifically, it should: (1) intensify packaging efforts; (2) commit additional resources to high-temperature electronics, including wide band-gap semiconductor devices such as SiC; and (3) redirect research on higher-speed electrical machines to improve torque density.

Structural Materials

Recommendation [3-17]. Based on the goal of 50 percent weight reduction as *a critical goal* and the near certainty that some (probably significant) cost penalty will be associated with it, the Partnership should develop a materials cost model (even if only an approximation) that can be used in a total systems model to spread this penalty in an optimal way across other vehicle components.

Recommendation [3-18]. The materials research funding should largely be redistributed to areas of higher potential payoff, such as high-energy batteries, fuel cells, hydrogen storage, and infrastructure issues. However, materials research for projects that show a high potential for enabling near-term, low-cost mass reduction should continue to be funded.

CHAPTER 4: HYDROGEN PRODUCTION, DELIVERY, AND DISPENSING

Hydrogen Fuel Pathways

Recommendation [4-1]. DOE should continue its studies of the transition to hydrogen, extending them to 2030-2035, a transition period during which the number of hydrogen vehicles in use could increase rapidly and use the results of these studies as a basis for evaluating the potential roles of different transitional supplies of hydrogen fuel as demand increases substantially, including both forecourt production at the fueling station and centralized production using the most cost-effective means of distributing the hydrogen.

Hydrogen Production

Recommendation [4-2]. DOE should conduct a systematic review of the CCS program as it affects the schedule for and program assumptions about hydrogen production from coal. This review should identify indicators of incipient program slippage and, through systems analysis, the program consequences of possible delays, leading to recommendations for management actions that would compensate for these delays.

Recommendation [4-3]. Like the hydrogen production from coal option, the Hydrogen, Fuel Cell and Infrastructure Technology (HFCIT) program should actively employ the liaison mechanisms put in place since the Phase 1 review. However, the exploratory nature of the programs for nuclear production suggests that, unlike the coal option, a detailed systems analysis of schedule delays would be premature at this time. Instead, systems analyses should focus on the complex interactions among program components, especially between the research elements of the nuclear and chemical processes, to ensure that technical progress in each distinct area leads ultimately to a practical system.

Recommendation [4-4]. The DOE should continue to promote electrolysis that uses renewable power integrated with electrolysis systems and to support analyses and demonstrations. High-temperature electrolysis activities within the Office of Nuclear Energy should be closely monitored.

Recommendation [4-5]. The Partnership should increase funding for electrolysis programs to advance the technology, demonstrations, and systems integration.

Recommendation [4-6]. Basic Energy Sciences should support, as appropriate, fundamental research in the area of catalysts, membranes, coatings, and new concepts.

Recommendation [4-7]. DOE should undertake a systems study to determine how best to use wind power–electrolysis combinations to generate hydrogen, considering overall cost and efficiency.

Recommendation [4-8]. The committee recommends that DOE projections of future hydrogen production for hydrogen-powered vehicles include scenarios in which the timetable for commercial quantities of these fuels is delayed, perhaps by as much as a decade.

Recommendation [4-9]. DOE should give priority to completing process development on biomass gasification, including any needed demonstration projects.

Recommendation [4-10]. DOE should undertake a systems study to assess the relative importance of barriers to biomass production, availability, transportation, and conversion to hydrogen; to identify the areas that are most important to commercial viability; and to give them priority. This study should address technical barriers already identified, including impact on the environment, and help define policies for land and water use and government-sponsored commercial incentives that would stimulate commercial expansion of the biomass options.

Recommendation [4-11]. DOE should involve the energy partners in all biomass programs related to conversion to hydrogen or hydrogen carriers as quickly as possible.

Recommendation [4-12]. Given the large number of potential ways of using biomass to supply hydrogen, DOE should identify the most promising approaches so it can focus on options that could have the greatest impact on hydrogen supply.

Recommendation [4-13]. DOE should put more emphasis on the space requirements for forecourt hydrogen generation by studying ways to minimize these requirements.

Hydrogen Delivery, Dispensing, and Transition Supply

Recommendation [4-14]. DOE should increase funding for the delivery and dispensing program to meet the market transition and sustained market penetration time frames. If DOE concludes that a funding increase is not feasible, the program should be focused on the pipeline, liquefaction, and compression programs, where a successful if only incremental short-term impact could be significant for the market transition period.

Recommendation [4-15]. DOE should, with the guidance of an independent outside committee, evaluate the achievability of the program's 2012 delivery and dispensing cost goal, \$1.00/kg H_2, particularly with 700 bar (10,000 psi) gas dispensing.

Recommendation [4-16]. DOE should consider supporting advanced systems engineering, integration, and demonstrations for home-based refueling systems, which should bring substantial learning value for such systems. This program should include careful consideration of operation and maintenance procedures that home owners are willing and able to perform.

Appendix E

Committee Meetings and Presentations

COMMITTEE MEETING, WASHINGTON, D.C., APRIL 27, 2009

See Appendix B, Committee's Interim Letter Report, Attachment III, for list of presentations.

COMMITTEE MEETING, SOUTHFIELD, MICHIGAN AUGUST 4-5, 2009

Introduction and Welcome
Vernon Roan, Committee Chair

Welcome
Don Walkowicz, Executive Director, U.S. Council for Automotive Research

Opening Remarks
Gerhard Schmidt, Ford

Automotive Perspective on the FreedomCAR and Fuel Program
John Sakioka, Ford

Fuel Perspective on the FreedomCAR and Fuel Program
Brad Smith, Shell

Utility Perspective on the FreedomCAR and Fuel Program
Knut Simonsen, DTE Energy

Overview of the FreedomCAR and Fuel Program
Pat Davis and Sunita Satyapal, U.S. Department of Energy (DOE)

Advanced Combustion and Emissions Control
Pete Moilanen, Ford
Ken Howden, DOE

Electrochemical Energy Storage
Kent Snyder, Ford
Dave Howell, DOE

Vehicle Systems Analysis
Larry Laws, General Motors (GM)
Steven Boyd, DOE

Electrical Systems and Electronics
John Czubay, GM
Susan Rogers, DOE

Grid Interaction Technical Team
Keith Hardy and Russ Conklin, DOE

Materials
Jim Quinn, GM
Joe Carpenter, DOE

Fuel Cells
Craig Gittleman, GM
Kathi Epping Martin, DOE

Onboard Hydrogen Storage
Andrea Sudik, Ford
Farshad Bavarian, Chevron
Ned Stetson, DOE

Codes and Standards
Mike Veenstra, Ford
Antonio Ruiz, DOE

Hydrogen Production
Nikunj Gupta, Shell
Roxanne Garland, DOE

Hydrogen Delivery
Jim Kegerreis, ExxonMobil
Monterey Gardiner, DOE

Fuel Pathway Integration
C.J. Guo, Shell
Fred Joseck, DOE

COMMITTEE MEETING, WASHINGTON, D.C. OCTOBER 26, 2009

FY 2010 Budget
Sunita Satyapal, DOE

Status and Outlook for Biofuels and Hydrogen from Biomass
Neil Rossmeissl, DOE

Status and Outlook for Hydrogen from Coal
Mark Ackiewicz, DOE

Status and Outlook for Carbon Capture and Storage
Lowell Miller, DOE

Hydrogen Storage Centers of Excellence
Ned Stetson, DOE

DOE'S Use of System Analysis
Fred Joseck, Lee Slezak, Pat Davis, and Sunita Satyapal, DOE

Status of 3M's Fuel Cell Efforts
M. Debe, 3M Company

DTI Fuel Cell Cost Analysis
B. James, DTI

Resource (Platinum and Natural Gas) Availability
Fred Joseck, DOE

Q&A: Batteries—PHEC and BEV Applications
Patrick Davis, DOE
J. Miller, Argonne National Laboratory

COMMITTEE MEETING, WASHINGTON, D.C.
DECEMBER 10, 2009

Evaluation of the Potential Environmental Impacts from Large-Scale Use and Production of Hydrogen in Energy and Transportation Applications
Don Wuebbles, University of Illinois at Urbana-Champaign

Potential Environmental Impacts of Hydrogen-Based Transportation and Power Systems
Tom Grieb, Tetra Tech, Inc.

Overview of U.S. Department of Transportation Hydrogen Activities
M.J. Fiocco, Research and Innovative Technology Administration, U.S. Department of Transportation

PHEV Update—U.S. Department of Energy Activities
David Howell, Vehicle Technologies Program, Energy Efficiency and Renewable Energy, DOE

The MA^3T Model: Market Acceptance of Advanced Automotive Technologies
David Greene, Oak Ridge National Laboratory
Zhenhong Lin, Oak Ridge National Laboratory

COMMITTEE SUBGROUP MEETING AT GENERAL MOTORS
HONEOYE FALLS, NEW YORK, SEPTEMBER 10, 2009

Discussions of General Motors' efforts on fuel cell vehicles.

Appendix F

Acronyms and Abbreviations

ac	alternating current
ACEC	advanced combustion and emission control (technical team)
ANL	Argonne National Laboratory
ARPA-E	Advanced Research Projects Agency-Energy (DOE)
ARRA	American Recovery and Reinvestment Act of 2009
bbl	barrel
BES	(Office of) Basic Energy Sciences (DOE)
BEV	battery electric vehicle
BGY	billion gallons per year
BoP	balance of plant
C&S	codes and standards
CAFE	corporate average fuel economy
CCS	carbon capture and storage
CEM	compressor expander motor
CFD	computational fluid dynamics
CLEERS	Crosscut Lean Exhaust Emission Reduction Simulation
CNG	compressed natural gas
CO	carbon monoxide
CO_2	carbon dioxide
COE	center of excellence
CRADA	collaborative research and development agreement
dc	direct current

DOE	U.S. Department of Energy
DOT	U.S. Department of Transportation
DPF	diesel particulate filter
E85	85 percent ethanol
EAC	Electricity Advisory Committee
EERE	(Office of) Energy Efficiency and Renewable Energy (DOE)
EGR	exhaust gas recirculation
EIA	U.S. Energy Information Administration
EISA	Energy Independence and Security Act of 2007
EPA	U.S. Environmental Protection Agency
EPRI	Electric Power Research Institute
ESG	Executive Steering Group
FACE	fuels for advanced combustion engines
FCFP	FreedomCAR and Fuel Partnership
FCHEV	fuel cell hybrid electric vehicle
FCT	Fuel Cell Technologies (program)
FCVT	FreedomCAR and Vehicle Technologies (program)
FE	(Office of) Fossil Energy (DOE)
FFV	flexible fuel vehicle
FMEA	failure modes and effects analysis
FPITT	fuel pathway integration technical team
FTA	Federal Transit Administration
FY	fiscal year
GaN	gallium nitride
gge	gallon gasoline equivalent
GHG	greenhouse gas
GREET	Greenhouse Gases, Regulated Emissions, and Energy Use in Transportation (model)
H, H_2	hydrogen
H2A	Hydrogen Technology (model)
HAMMER	Hazardous Materials Management and Emergency Response (facility)
HC	hydrocarbon
HCCI	homogeneous charge compression ignition
HEV	hybrid electric vehicle
HFCIT	Hydrogen, Fuel Cells and Infrastructure Technologies (program)
HFCV	hydrogen fuel cell vehicle
HFI	Hydrogen Fuels Initiative
HHV	higher heating value

HyTrans	Hydrogen Transition (model)
ICE	internal combustion engine
IEEE	Institute for Electrical and Electronics Engineers
IMEP	indicated mean effective pressure
IPM	interior permanent magnet
ISO	International Organization for Standardization
kg	kilogram
kW	kilowatt
kWe	kilowatt (electric)
kWh	kilowatt-hour
Li-ion	lithium-ion
LCA	life-cycle assessment
LDV	light-duty vehicle
LHV	lower heating value
LPG	liquefied petroleum gas
LTC	low-temperature combustion
MARKAL	Market Analysis (model)
MEA	membrane electrode assembly
MFA	materials flow analysis
MOU	memorandum of understanding
MPa	megapascal
mpg	miles per gallon
MSM	Macro System Model
MT	metric ton
NAE	National Academy of Engineering
NAS	National Academy of Sciences
NE	(Office of) Nuclear Energy (DOE)
NEMS	National Energy Modeling System
NERC	North American Electric Reliability Corporation
NETL	National Energy Technology Laboratory
NFPA	National Fire Protection Association
NGNP	Next Generation Nuclear Powerplant
NHTSA	National Highway Traffic Safety Administration
NiMH	nickel metal hydride
NIST	National Institute of Standards and Technology
NO_x	nitrogen oxides
NPC	National Petroleum Council
NPV	net present value

NRC	National Research Council
NRDC	Natural Resources Defense Council
NREL	National Renewable Energy Laboratory
OEM	original equipment manufacturer
ORNL	Oak Ridge National Laboratory
PAN	polyacrylonitrile
PbA	lead acid (battery)
PEM	proton exchange membrane
PHEV	plug-in hybrid electric vehicle
PHMSA	Pipeline and Hazardous Materials Safety Administration
PM	particulate matter
PNGV	Partnership for a New Generation of Vehicles
PNNL	Pacific Northwest National Laboratory
PRD	pressure relief device
PSAT	Powertrain Systems Analysis Toolkit
R&D	research and development
SAE	Society of Automotive Engineers
SBIR	Small Business Innovation Research
SC	(Office of) Science (DOE)
SCR	selective catalytic reduction
SiC	silicon carbide
SNL	Sandia National Laboratories
SOC	state of charge
SOI	silicon on insulator
SRNL	Savannah River National Laboratory
STTR	small business technology transfer
SUV	sport utility vehicle
21CTP	21st Century Truck Partnership
UPS	uninterruptible power supply
USABC	United States Advanced Battery Consortium
USCAR	U.S. Council for Automotive Research
VSATT	vehicle systems analysis technical team
VT	Vehicle Technologies (Office of)